Nanotechnology: Tools and Techniques

Nanotechnology
Tools and Techniques

Anandhakrishna Rao

ADHYAYAN

ADHYAYAN PUBLISHERS & DISTRIBUTORS
New Delhi-110002

Published by
ADHYAYAN PUBLISHERS & DISTRIBUTORS
4378/4B, 105, J.M.D. House, Murari Lal Street
Ansari Road, Darya Ganj, New Delhi - 110002
Ph.: 011-23263018, 011-23277156, Fax: 011-23280028
Email: adhyayanpublishers@yahoo.com

Nanotechnology: Tools and Techniques

Edition 2010
ISBN 978-81-8435-185-9

Printed in India

Published by Harish Chandra Yadav for Adhyayan Publishers & Distributors, Delhi. Laser Typesetting at V. M. Graphics and Printed at Tarun Offset Printers, Delhi

Preface

Nanotechnology is dealing with research and the construction in the sphere of very small structures: one nanometre is one millionth of a millimetre. Applications emerge in energy technology, in environmental technology or in information technology but also in the healthcare area. Nanotechnology is the umbrella term for the most different types of analysis and processing of materials which have one thing in common: Their size is one to one hundred nanometres. Nanotechnology makes use of the special characteristics of many nanostructures. The mechanical, optical, magnetic, electrical and chemical characteristics of these very small structures do not only depend on their original material but very much also on their size and shape.

Nanotechnology develops the basis for increasingly smaller data memories with increasingly larger storage capacity, for highly efficient filters for sewage treatment, for photovoltaic windows, for materials which can be used to build ultra-light engines and body parts in the automobile industry or for artificial joints which are better tolerated by the human body due to organic nano-surfaces. Nanotechnology is increasingly considered to be the future technology. Instead of "ever higher, ever wider" its motto is "ever smaller, ever faster". Nanotechnology provides access to the world of the smallest things. The possible applications of this technology are immense. Future progress in nanotechnology will also determine the further development of future-oriented branches.

This book is an authoritative, indepth exploration of the tools and techniques of nanotechnology in a lucid style. It will make fascinating reading for engineers, scientists, administrators and students in a range of science and engineering disciplines.

Anandhakrishna Rao

Contents

1

Introduction

The first use of the concepts in 'nano-technology' was in "There's Plenty of Room at the Bottom," a talk given by physicist Richard Feynman at an American Physical Society meeting at Caltech on December 29, 1959. Feynman described a process by which the ability to manipulate individual atoms and molecules might be developed, using one set of precise tools to build and operate another proportionally smaller set, so on down to the needed scale. In the course of this, he noted, scaling issues would arise from the changing magnitude of various physical phenomena: gravity would become less important, surface tension and Van der Waals attraction would become more important, etc. This basic idea appears plausible, and exponential assembly enhances it with parallelism to produce a useful quantity of end products.

The term "nanotechnology" was defined by Tokyo Science University Professor Norio Taniguchi in a 1974 paper as follows: "'Nano-technology' mainly consists of the processing of, separation, consolidation, and deformation of materials by one atom or by one molecule." In the 1980s the basic idea of this definition was explored in much more depth by Dr. K. Eric Drexler, who promoted the technological significance of nano-scale phenomena and devices through speeches and the books Engines of Creation: The Coming Era of Nanotechnology and Nanosystems: Molecular Machinery, Manufacturing, and Computation, and so the term acquired its current sense.

The Coming Era of Nanotechnology is considered the first book on the topic of nanotechnology. Nanotechnology and nanoscience got started in the early 1980's with two major developments; the birth of cluster science and the invention of the scanning tunnelling microscope (STM). This development led to the discovery of fullerenes in 1986 and carbon nanotubes a few years later.

In another development, the synthesis and properties of semiconductor nanocrystals was studied; This led to a fast increasing number of metal oxide nanoparticles of quantum dots. The atomic force microscope was invented six years after the STM was invented. In 2000, the United States National Nanotechnology Initiative was founded to coordinate Federal nanotechnology research and development.

Humans have unwittingly employed nanotechnology for thousands of years, for example in making steel, paintings and in vulcanising rubber. Each of these processes rely on the properties of stochastically-formed atomic ensembles mere nanometers in size, and are distinguished from chemistry in that they don't rely on the properties of individual molecules. But the development of the body of concepts now subsumed under the term nanotechnology has been slower.

The first mention of some of the distinguishing concepts in nanotechnology was in 1867 by James Clerk Maxwell when he proposed as a thought experiment a tiny entity known as Maxwell's Demon able to handle individual molecules. The first observations and size measurements of nano-particles was made during first decade of 20th century.

They are mostly associated with Richard Adolf Zsigmondy who made detail study of gold sols and other nanomaterials with sizes down to 10 nm and less. He published a book in 1914. He used ultramicroscope that employes dark field method for seeing particles with sizes much less than light wavelength. Zsigmondy was also the first who used nanometer explicitly for characterising particle size. He determined it as 1/1,000,000 of millimetre. He developed a first system classification based on particle size in nanometer range.

There have been many significant developments during 20th century in characterising nanomaterials and related phenomena.

belonging to the field of interface and colloid science. In the 1920s, Irving Langmuir and Katharine B. Blodgett introduced the concept of a monolayer, a layer of material one molecule thick. Langmuir won a Nobel Prize in chemistry for his work. In early 1950s, Derjaguin and Abrikosova conducted the first measurement of surface forces.

There have been many studies of periodic colloidal structures and principles of molecular self-assembly that are overviewed in the paper. There are many other discoveries that serve as scientific basis for the modern nanotechnology can be found in the Fundamentals of Interface and Colloid Science by H.Lyklema.

Conceptual Origins

The topic of nanotechnology was again touched upon by "There's Plenty of Room at the Bottom," a talk given by physicist Richard Feynman at an American Physical Society meeting at Caltech on December 29, 1959. Feynman described a process by which the ability to manipulate individual atoms and molecules might be developed, using one set of precise tools to build and operate another proportionally smaller set, so on down to the needed scale.

In the course of this, scaling issues would arise from the changing magnitude of various physical phenomena: gravity would become less important, surface tension and Van der Waals attraction would become more important, etc. This basic idea appears feasible, and exponential assembly enhances it with parallelism to produce a useful quantity of end products. At the meeting, Feynman announced two challenges, and he offered a prize of $1000 for the first individuals to solve each one.

The first challenge involved the construction of a nanomotor, which, to Feynman's surprise, was achieved by November of 1960 by William McLellan. The second challenge involved the possibility of scaling down letters small enough so as to be able to fit the entire Encyclopedia Britannica on the head of a pin; this prize was claimed in 1985 by Tom Newman.

In 1965 Gordon Moore observed that silicon transistors were undergoing a continual process of scaling downward, an observation which was later codified as Moore's law. Since his observation transistor minimum feature sizes have decreased from 10 micrometers to the 45-65 nm range in 2007; one minimum feature is thus roughly 180 silicon atoms long.

The term "nanotechnology" was first defined by Norio Taniguchi of the Tokyo Science University in a 1974 paper as follows: "'Nano-technology' mainly consists of the processing of, separation, consolidation, and deformation of materials by one atom or one molecule." Since that time the definition of nanotechnology has generally been extended to include features as large as 100 nm. Additionally, the idea that nanotechnology embraces structures exhibiting quantum mechanical aspects, such as quantum dots, has further evolved its definition.

Also in 1974 the process of atomic layer deposition, for depositing uniform thin films one atomic layer at a time, was developed and patented by Dr. Tuomo Suntola and co-workers in Finland. In the 1980s the idea of nanotechnology as deterministic, rather than stochastic, handling of individual atoms and molecules was conceptually explored in depth by Dr. K. Eric Drexler, who promoted the technological significance of nano-scale phenomena and devices through speeches and the books Engines of Creation: The Coming Era of Nanotechnology and Nanosystems: Molecular Machinery, Manufacturing, and Computation. Drexler's vision of nanotechnology is often called "Molecular Nanotechnology" (MNT) or "molecular manufacturing," and Drexler at one point proposed the term "zettatech" which never became popular.

Experimental Advances

Nanotechnology and nanoscience got a boost in the early 1980s with two major developments: the birth of cluster science and the invention of the scanning tunnelling microscope (STM). This development led to the discovery of fullerenes in 1985 and the structural assignment of carbon nanotubes a few years later. In another development, the synthesis and properties of semiconductor nanocrystals were studied. This led to a fast increasing number of semiconductor nanoparticles of quantum dots.

In the early 1990s Huffman and Kraetschmer, of the University of Arizona, discovered how to synthesize and purify large quantities of fullerenes. This opened the door to their characterisation and functionalisation by hundreds of investigators in government and industrial laboratories.

Rubidium doped C60 was found to be a mid temperature (Tc = 32 K) superconductor. At a meeting of the Materials Research Society in 1992, Dr. T. Ebbesen (NEC) described to a spellbound audience his discovery and characterisation of carbon nanotubes. This event sent those in attendance and others downwind of his presentation into their laboratories to reproduce and push those discoveries forward. Using the same or similar tools as those used by Huffman and Kratschmere, hundreds of researchers further developed the field of nanotube-based nanotechnology.

At present in 2007 the practice of nanotechnology embraces both stochastic approaches and deterministic approaches wherein single molecules are manipulated on substrate surfaces by deterministic methods comprising nudging them with STM or AFM probes and causing simple binding or cleavage reactions to occur. The dream of a complex, deterministic molecular nanotechnology remains elusive.

Since the mid 1990s, thousands of surface scientists and thin film technocrats have latched on to the nanotechnology bandwagon and redefined their disciplines as nanotechnology. This has caused much confusion in the field and has spawned thousands of "nano"-papers on the peer reviewed literature. Most of these reports are extensions of the more ordinary research done in the parent fields.

For the future, some means has to be found for MNT design evolution at the nanoscale which mimics the process of biological evolution at the molecular scale. Biological evolution proceeds by random variation in ensemble averages of organisms combined with culling of the less-successful variants and reproduction of the more-successful variants, and macroscale engineering design also proceeds by a process of design evolution from simplicity to complexity as set forth somewhat satirically by John Gall: "A complex system that works is invariably found to have evolved from a simple system that worked.... A complex system designed from scratch never works and can not be patched up to make it work. You have to start over, beginning with a system that works."

A breakthrough in MNT is needed which proceeds from the simple atomic ensembles which can be built with, e.g., an STM to complex MNT systems via a process of design evolution. A handicap in this process is the difficulty of seeing and manipulation at the nanoscale compared to the macroscale which makes deterministic selection of successful trials difficult; in contrast biological evolution proceeds via action of what Richard Dawkins has called the "blind watchmaker" comprising random molecular variation and deterministic reproduction/extinction.

Define Nanomaterials

Nanomaterials are applications with morphological features smaller than a one tenth of a micrometre in at least one dimension. Despite the fact that there is no consensus upon the minimum or maximum size of nanomaterials, with some authors restricting their size to as low as 1 to ~30 nm, a logical definition would situate the nanoscale between microscale (0.1 micrometre) and atomic/molecular scale (about 0.2 nanometers).

Concepts of Nanomaterials

An aspect of nanotechnology is the vastly increased ratio of surface area to volume present in many nanoscale materials which makes possible new quantum mechanical effects, for example the "quantum size effect" where the electronic properties of solids are altered with great reductions in particle size. This effect does not come into play by going from macro to micro dimensions. However, it becomes pronounced when the nanometer size range is reached. A certain number of physical properties also alter with the change from macroscopic systems.

Novel mechanical properties of nanomaterials is a subject of nanomechanics research. Catalytic activities also reveal new behaviour in the interaction with biomaterials. Nanotechnology can be thought of as extensions of traditional disciplines towards the explicit consideration of these properties. Additionally, traditional disciplines can be reinterpreted as specific applications of nanotechnology. This dynamic reciprocation of ideas and concepts contributes to the modern understanding of the field.

Broadly speaking, nanotechnology is the synthesis and application of ideas from science and engineering towards the understanding and production of novel materials and devices. These products generally make copious use of physical properties associated with small scales. Materials reduced to the nanoscale can suddenly show very different properties compared to what they exhibit on a macroscale, enabling unique applications.

For instance, opaque substances become transparent (copper); inert materials attain catalytic properties (platinum); stable materials turn combustible (aluminium); solids turn into liquids at room temperature (gold); insulators become conductors (silicon). Materials such as gold, which is chemically inert at normal scales, can serve as a potent chemical catalyst at nanoscales. Much of the fascination with nanotechnology stems from these unique quantum and surface phenomena that matter exhibits at the nanoscale.

Nanosize powder particles are potentially important in ceramics, powder metallurgy, the achievement of uniform nanoporosity and similar applications. The strong tendency of small particles to form clumps is a serious technological problem that impedes such applications. However, a number of dispersants such as ammonium citrate and imidazoline or oleyl alcohol are promising solutions as possible additives for deagglomeration.

Size Concerns

Another concern is that the volume of an object decreases as the third power of its linear dimensions, but the surface area only decreases as its second power. This somewhat subtle and unavoidable principle has huge ramifications. For example the power of a drill is proportional to the volume, while the friction of the drill's bearings and gears is proportional to their surface area.

For a normal-sized drill, the power of the device is enough to handily overcome any friction. However, scaling its length down by a factor of 1000, for example, decreases its power by 10003 (a factor of a billion) while reducing the friction by only 10002. Proportionally it has 1000 times less power per unit friction than the original drill. If the original friction-to-power ratio was, say, 1%, that implies the smaller drill will have 10 times as much friction as power. The drill is useless.

For this reason, while super-miniature electronic integrated circuits are fully functional, the same technology cannot be used to make working mechanical devices beyond the scales where frictional forces start to exceed the available power. So even though you may see microphotographs of delicately etched silicon gears, such devices are currently little more than curiosities with limited real world applications, for example in moving mirrors and shutters.

Surface tension increases in much the same way, thus magnifying the tendency for very small objects to stick together. This could possibly make any kind of "micro factory" impractical: even if robotic arms and hands could be scaled down, anything they pick up will tend to be impossible to put down.

Molecular evolution has resulted in working cilia, flagella, muscle fibers and rotary motors in aqueous environments, all on the nanoscale. These machines exploit the increased frictional forces found at the micro or nanoscale. Unlike a paddle or a propeller which depends on normal frictional forces to achieve propulsion, cilia develop motion from the exaggerated drag or laminar forces present at micro and nano dimensions. To build meaningful "machines" at the nanoscale, the relevant forces need to be considered.

Nanomaterials

Materials referred to as "nanomaterials" generally fall into two categories: fullerenes, and inorganic nanoparticles.

Fullerenes

The fullerenes are a class of allotropes of carbon which conceptually are graphene sheets rolled into tubes or spheres. These include the carbon nanotubes which are of interest both because of their mechanical strength and also because of their electrical properties. For the past decade, the chemical and physical properties of fullerenes have been a hot topic in the field of research and development, and are likely to continue to be for a long time.

In April 2003, fullerenes were under study for potential medicinal use: binding specific antibiotics to the structure of resistant bacteria and even target certain types of cancer cells such as melanoma.

The October 2005 issue of Chemistry and Biology contains an article describing the use of fullerenes as light-activated antimicrobial agents. In the field of nanotechnology, heat resistance and superconductivity are among the properties attracting intense research.

A common method used to produce fullerenes is to send a large current between two nearby graphite electrodes in an inert atmosphere. The resulting carbon plasma arc between the electrodes cools into sooty residue from which many fullerenes can be isolated. There are many calculations that have been done using ab-initio Quantum Methods applied to fullerenes. By DFT and TDDFT methods one can obtain IR, Raman and UV spectra. Results of such calculations can be compared with experimental results.

Effects of Nanoparticles

Nanoparticles or nanocrystals made of metals, semiconductors, or oxides are of particular interest for their mechanical, electrical, magnetic, optical, chemical and other properties. Nanoparticles have been used as quantum dots and as chemical catalysts. Nanoparticles are of great scientific interest as they are effectively a bridge between bulk materials and atomic or molecular structures. A bulk material should have constant physical properties regardless of its size, but at the nano-scale this is often not the case.

Size-dependent properties are observed such as quantum confinement in semiconductor particles, surface plasmon resonance in some metal particles and superparamagnetism in magnetic materials. Nanoparticles exhibit a number of special properties relative to bulk material. For example, the bending of bulk copper occurs with movement of copper atoms/clusters at about the 50 nm scale. Copper nanoparticles smaller than 50 nm are considered super hard materials that do not exhibit the same malleability and ductility as bulk copper.

The change in properties is not always desirable. Ferroelectric materials smaller than 10 nm can switch their magnetisation direction using room temperature thermal energy, thus making them useless for memory storage. Suspensions of nanoparticles are possible because the interaction of the particle surface with the solvent is strong enough to overcome differences in density, which usually result in a material either sinking or floating in a liquid. Nanoparticles often have unexpected visual properties because they are small enough to confine their electrons and produce quantum effects.

The often very high surface area to volume ratio of nanoparticles provides a tremendous driving force for diffusion, especially at elevated temperatures. Sintering is possible at lower temperatures and over shorter durations than for larger particles. This theoretically does not affect the density of the final product, though flow difficulties and the tendency of nanoparticles to agglomerate do complicate matters. The surface effects of nanoparticles also reduces the incipient melting temperature.

Manufactured Nanomaterials

Nanomaterials behave differently than other similarly-sized particles. It is therefore necessary to develop specialised approaches to testing and monitoring their effects on human health and on the environment. The OECD Chemicals Committee has established the Working Party on Manufactured Nanomaterials to address this issue and to study the practices of OECD member countries in regards to nanomaterial safety.

While nanomaterials and nanotechnologies are expected to yield numerous health and health care advances, such as more targeted methods of delivering drugs, new cancer therapies, and methods of early detection of diseases, they also may have unwanted effects. Increased rate of absorption is the main concern associated with manufactured nanoparticles. When materials are made into nanoparticles, their surface area to volume ratio increases.

The greater specific surface area may lead to increased rate of absorption through the skin, lungs, or digestive tract and may cause unwanted effects to the lungs as well as other organs. However, the particles must be absorbed in sufficient quantities in order to pose health risks. As the use of nanomaterials increases worldwide, concerns for worker and user safety are mounting. To address such concerns, the Swedish Karolinska Institute conducted a study in which various nanoparticles were introduced to human lung epithelial cells.

The results, released in 2008, showed that iron oxide nanoparticles caused little DNA damage and were non-toxic. Zinc oxide nanoparticles were slightly worse. Titanium dioxide caused only DNA damage. Carbon nanotubes caused DNA damage at low levels. Copper oxide was found to be the worst offender, and was the only nanomaterial identified by the researchers as a clear health risk.

MOLECULAR SELF-ASSEMBLY

Molecular self-assembly is the process by which molecules adopt a defined arrangement without guidance or management from an outside source. There are two types of self-assembly, intramolecular self-assembly and intermolecular self-assembly. Most often the term molecular self-assembly refers to intermolecular self-assembly, while the intramolecular analog is more commonly called folding.

Supramolecular Systems

Molecular self-assembly is a key concept in supramolecular chemistry since assembly of the molecules is directed through noncovalent interactions. Common examples include the formation of micelles, vesicles, liquid crystal phases, and Langmuir monolayers by surfactant molecules. Molecular self-assembly has allowed the construction of challenging molecular topologies. An example are Borromean rings, interlocking rings wherein removal of one ring unlocks each of the other rings. DNA has been used to prepare a molecular analog of Borromean rings.

Biological Systems

Molecular self-assembly is crucial to the function of cells. It is exhibited in the self-assembly of lipids to form the membrane, the formation of double helical DNA through hydrogen bonding of the individual strands, and the assembly of proteins to form quaternary structures. Molecular self-assembly of incorrectly folded proteins into insoluble amyloid fibers is responsible for infectious prion-related neurodegenerative diseases.

Nanotechnology

Molecular self-assembly is an important aspect of bottom-up

approaches to nanotechnology. Using molecular self-assembly the final (desired) structure is programmed in the shape and functional groups of the molecules. Self-assembly is referred to as a 'bottom-up' manufacturing technique in contrast to a 'top-down' technique such as lithography where the desired final structure is carved from a larger block of matter.

In the speculative vision of molecular nanotechnology, microchips of the future might be made by molecular self-assembly. An advantage to constructing nanostructure using molecular self-assembly for biological materials is that they will degrade back into individual molecules that can be broken down by the body.

DNA nanotechnology

DNA nanotechnology is an area of current research that uses the bottom-up, self-assembly approach for nanotechnological goals. DNA nanotechnology uses the unique molecular recognition properties of DNA and other nucleic acids to create self-assembling branched DNA complexes with useful properties.

DNA is thus used as a structural material rather than as a carrier of biological information, to make structures such as two-dimensional periodic lattices (both tile-based as well as using the "DNA origami" method) and three-dimensional structures in the shapes of polyhedra. These DNA structures have also been used to template the assembly of other molecules such as gold nanoparticles and streptavidin proteins.

Molecular Nanotechnology

Molecular nanotechnology (MNT) is the concept of engineering functional mechanical systems at the molecular scale. An equivalent definition would be "machines at the molecular scale designed and built atom-by-atom". This is distinct from nanoscale materials. Based on Richard Feynman's vision of miniature factories using nanomachines to build complex products (including additional nanomachines), this advanced form of nanotechnology would make use of positionally-controlled mechanosynthesis guided by molecular machine systems. MNT would involve combining physical principles demonstrated by chemistry, other nanotechnologies, and the molecular machinery of life with the systems engineering principles found in modern macroscale factories.

While conventional chemistry uses inexact processes driven toward some balance to obtain inexact results, and biology exploits inexact processes to obtain definitive results, molecular nanotechnology would employ original definitive processes to obtain definitive results. The desire in molecular nanotechnology would be to balance molecular reactions in positionally-controlled locations and orientations to obtain desired chemical reactions, and then to build systems by further assembling the products of these reactions.

Formulating a roadmap for the development of MNT is now an objective of a broadly based technology roadmap project led by Battelle and the Foresight Institute. The roadmap was originally scheduled for completion by late 2006, then by early 2007, and most recently is set to be unveiled in October 2007.

The Nanofactory Collaboration is a more focused ongoing effort involving 23 researchers from 10 organisations and 4 countries that is developing a practical research agenda specifically aimed at positionally-controlled diamond mechanosynthesis and diamondoid nanofactory development. In August 2005, a task force consisting of 50+ international experts from various fields was organised by the Center for Responsible Nanotechnology to study the societal implications of molecular nanotechnology.

Projected Applications

Smart materials

One proposed application of MNT is the development of so-called smart materials. This term refers to any sort of material designed and engineered at the nanometer scale to perform a specific task, and encompasses a wide variety of possible commercial applications. One example would be materials designed to respond differently to various molecules; such a capability could lead, for example, to artificial drugs which would recognise and render inert specific viruses.

Another is the idea of self-healing structures, which would repair small tears in a surface naturally in the same way as self-sealing tires or human skin. A nanosensor created by MNT would resemble a smart material, involving a small component within a larger machine that would react to its environment and change in some fundamental, intentional way.

As a very simple example: a photosensor could passively measure the incident light and discharge its absorbed energy as electricity when the light passes above or below a specified threshold, sending a signal to a larger machine. Such a sensor would supposedly cost less and use less power than a conventional sensor, and yet function usefully in all the same applications—for example, turning on parking lot lights when it gets dark. While smart materials and nanosensors both exemplify useful applications of MNT, they pale in comparison with the complexity of the technology most popularly associated with the term: the replicating nanorobot.

Replicating nanorobots

MNT nanofacturing is popularly linked with the idea of swarms of coordinated nanoscale robots working together, a popularisation of an early proposal by Drexler in his 1986 discussions of MNT, but superseded in 1992. In this early proposal, sufficiently capable nanorobots would construct more nanorobots in an artificial environment containing special molecular building blocks.

Critics have doubted both the feasibility of self-replicating nanorobots and the feasibility of control if self-replicating nanorobots could be achieved: they cite the possibility of mutations removing any control and favouring reproduction of mutant pathogenic variations.

Advocates address the first doubt by pointing out that the first macroscale autonomous machine replicator, made of Lego blocks, was built and operated experimentally in 2002. While there are sensory advantages present at the macroscale compared to the limited sensorium available at the nanoscale, proposals for positionally controlled nanoscale mechanosynthetic fabrication systems employ dead reckoning of tooltips combined with reliable reaction sequence design to ensure reliable results, hence a limited sensorium is no handicap; similar considerations apply to the positional assembly of small nanoparts.

Advocates address the second doubt by arguing that bacteria are evolved to evolve, while nanorobot mutation could be actively prevented by common error-correcting techniques. Similar ideas are advocated in the Foresight Guidelines on Molecular Nanotechnology, and a map of the 137-dimensional replicator design space recently published by Freitas and Merkle provides numerous proposed methods by which replicators could, in principle, be safely controlled by good design.

However, the concept of suppressing mutation raises the question: How can design evolution occur at the nanoscale without a process of random mutation and deterministic selection? Critics argue that MNT advocates have not provided a substitute for such a process of evolution in this nanoscale arena where conventional sensory-based selection processes are lacking. The limits of the sensorium available at the nanoscale could make it difficult or impossible to winnow successes from failures.

Advocates argue that design evolution should occur deterministically and strictly under human control, using the conventional engineering paradigm of modelling, design, prototyping, testing, analysis, and redesign. The limited sensorium is no handicap because, for example, prototype nanoparts could be fabricated via dead reckoning using positionally controlled chemically active tooltips, then characterised by chemically inactive scanning probe tooltips or other technical means, with errors corrected or design changes implemented in the next prototyping iteration.

Medical nanorobots

One of the most important applications of MNT would be medical nanorobotics or nanomedicine, an area pioneered by Robert Freitas in numerous books and papers. The ability to design, build, and deploy large numbers of medical nanorobots would, at an optimum, make possible the rapid elimination of disease and the reliable and relatively painless recovery from physical trauma. Medical nanorobots might also make possible the convenient correction of genetic defects, and help to ensure a greatly expanded healthspan.

More controversially, medical nanorobots might be used to augment natural human capabilities. However, mechanical medical nanodevices would not be allowed (or designed) to self-replicate inside the human body, nor would medical nanorobots have any need for self-replication themselves since they would be manufactured exclusively in carefully regulated nanofactories.

Utility fog

Another proposed application of molecular nanotechnology is "utility fog"- in which a cloud of networked microscopic robots (simpler than assemblers) would change its shape and properties to form macroscopic objects and tools in accordance with software commands.

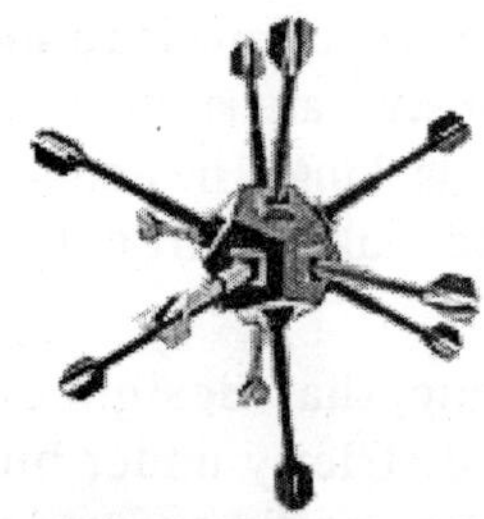

Diagram of a 100 micrometer foglet

Rather than modify the current practices of consuming material goods in different forms, utility fog would simply replace many physical objects.

Phased-array optics

Yet another proposed application of MNT would be phased-array optics (PAO). However, this appears to be a problem addressable by ordinary nanoscale technology. PAO would use the principle of phased-array millimetre technology but at optical wavelengths. This would permit the duplication of any sort of optical effect but virtually. Users could request holograms, sunrises and sunsets, or floating lasers as the mood strikes. PAO systems were described in BC Crandall's Nanotechnology: Molecular Speculations on Global Abundance in the Brian Wowk article "Phased-Array Optics."

Potential Social Impacts

Despite the current early developmental status of nanotechnology and molecular nanotechnology, much concern surrounds MNT's anticipated impact on economics and on law. Some conjecture that MNT would elicit a strong public-opinion backlash, as has occurred recently around genetically modified plants and the prospect of human cloning. Whatever the exact effects, MNT, if achieved, would tend to upset existing economic structures by reducing the scarcity of manufactured goods and making many more goods (such as food and health aids) manufacturable.

It is generally considered that future citizens of a molecular-nanotechnological society would still need money, in the form of unforgeable digital cash or physical specie. They might use such money to buy goods and services that are unique, or limited within the solar system. These might include: matter, energy, information, real estate, design services, entertainment services, legal services, fame, political power, or the attention of other people to your political/religious/philosophical message. Furthermore, futurists must consider war, even between prosperous states, and non-economic goals.

If MNT were realised, some resources would remain limited, because unique physical objects are limited or because they depend on the goodwill of a particular person. Demand will always exceed supply for some things, and a political economy may continue to exist in any case. Whether the interest in these limited resources would diminish with the advent of virtual reality, where they could be easily substituted, is yet unclear; one reason why it might not is a hypothetical preference for "the real thing".

Molecular nanotechnology also raises the feasibility of repairing cells that have been vitrified through cryonic preservation as well as abolishing the diseases which are not curable by today's means when an individual is reborn from cryonic stasis—creating conditions whereby individuals could be held accountable for their actions after death.

Risks

Molecular nanotechnology is one of the technologies that some analysts believe could lead to a Technological Singularity. Some feel that molecular nanotechnology would have daunting risks. It conceivably could enable cheaper and more destructive conventional weapons. Also, molecular nanotechnology might permit weapons of mass destruction that could self-replicate, as viruses and cancer cells do when attacking the human body. Commentators generally agree that, in the event molecular nanotechnology were developed, mankind should permit self-replication only under very controlled or "inherently safe" conditions.

A fear exists that nanomechanical robots, if achieved, and if designed to self-replicate using naturally occurring materials, could consume the entire planet in their hunger for raw materials, or simply crowd out natural life, out-competing it for energy. Some commentators have referred to this situation as the "grey goo" or "ecophagy" scenario. K. Eric Drexler considers an accidental "grey goo" scenario extremely unlikely and says so in later editions of Engines of Creation. The "grey goo" scenario begs the Tree Sap Answer: what chances exist that one's car could spontaneously mutate into a wild car, run off-road and live in the forest off tree sap?

In light of this perception of potential danger, the Foresight Institute has drafted a set of guidelines for the ethical development of nanotechnology. These include the banning of free-foraging self-replicating pseudo-organisms on the Earth's surface, at least, and possibly in other places.

Technical Issues and Criticism

Universal assemblers

A section heading in Drexler's Engines of Creation reads "Universal Assemblers", and the following text speaks of molecular assemblers which could hypothetically "build almost anything that the laws of nature allow to exist." Drexler's colleague Ralph Merkle has noted that, contrary to widespread legend, Drexler never claimed that assembler systems could build absolutely any molecular structure.

The endnotes in Drexler's book explain the qualification "almost": "For example, a delicate structure might be designed that, like a stone arch, would self-destruct unless all its pieces were already in place. If there were no room in the design for the placement and removal of a scaffolding, then the structure might be impossible to build. Few structures of practical interest seem likely to exhibit such a problem, however."

In 1992, Drexler published Nanosystems: Molecular Machinery, Manufacturing, and Computation, a detailed proposal for synthesizing stiff covalent structures using a table-top factory. Although such a nanofactory would be far less powerful than a protean universal assembler, it would still be enormously capable.

Diamondoid structures and other stiff covalent structures, if achieved, would have a wide range of possible applications, going far beyond current MEMS technology. An outline of a path was put forward in 1992 for building a table-top factory in the absence of a near-universal assembler. Other researchers have begun advancing tentative, alternative proposed paths for this in the years since Nanosystems was published.

Smalley-Drexler debate

Several researchers, including Nobel Prize winner Dr. Richard Smalley, attacked the notion of universal assemblers, leading to a rebuttal from Drexler and colleagues, and eventually to an exchange of letters. Smalley argued that chemistry is extremely complicated, reactions are hard to control, and that a universal assembler is science fiction. Drexler and colleagues, however, noted that Drexler never proposed universal assemblers able to make absolutely anything, but instead proposed more limited assemblers able to make a very wide variety of things. They challenged the relevance of Smalley's arguments to the more specific proposals advanced in Nanosystems.

Proposals in nanosystems

The feasibility of Drexler's proposals largely depends, therefore, on whether designs like those in Nanosystems could be built in the absence of a universal assembler to build them and would work as described. Supporters of molecular nanotechnology frequently claim that no significant errors have been discovered in Nanosystems since 1992. Even some critics concede that "Drexler has carefully considered a number of physical principles underlying the 'high level' aspects of the nanosystems he proposes and, indeed, has thought in some detail" about some issues.

Other critics claim, however, that Nanosystems omits important chemical details about the low-level 'machine language' of molecular nanotechnology. They also claim that much of the other low-level chemistry in Nanosystems requires extensive further work, and that Drexler's higher-level designs therefore rest on speculative foundations. Recent such further work by Freitas and Merkle is aimed at strengthening these foundations by filling the existing gaps in the low-level chemistry.

Drexler argues that we may need to wait until the conventional nanotechnology improves before solving these issues: "Molecular manufacturing will result from a series of advances in molecular machine systems, much as the first Moon landing resulted from a series of advances in liquid-fuel rocket systems. We are now in a position like that of the British Interplanetary Society of the 1930s which described how multistage liquid-fuelled rockets could reach the Moon and pointed to early rockets as illustrations of the basic principle."

However, Freitas and Merkle argue that a focused effort to achieve diamond mechanosynthesis (DMS) can begin now, using existing technology, and might achieve success in less than a decade if their "direct-to-DMS approach is pursued rather than a more circuitous development approach that seeks to implement less efficacious nondiamondoid molecular manufacturing technologies before progressing to diamondoid".

To summarise the arguments against feasibility: First, critics argue that a primary barrier to achieving molecular nanotechnology is the lack of an efficient way to create machines on a molecular/atomic scale, especially in the absence of a well-defined path toward a self-replicating assembler or diamondoid nanofactory. Advocates respond that a preliminary research path leading to a diamondoid nanofactory is being developed.

A second difficulty in reaching molecular nanotechnology is design. Hand design of a gear or bearing at the level of atoms is a gruelling task. While Drexler, Merkle and others have created a few designs of simple parts, no comprehensive design effort for anything approaching the complexity of a Model T Ford has been attempted. Advocates respond that it is difficult to undertake a comprehensive design effort in the absence of significant funding for such efforts, and that despite this handicap much useful design-ahead has nevertheless been accomplished with new software tools that have been developed, e.g., at Nanorex.

A third difficulty in achieving molecular technology is separating successful trials from failures, and elucidating the failure mechanisms of the failures. Unlike biological evolution, which proceeds by random variations in ensembles of organisms combined with deterministic reproduction/extinction as a selection process to achieve great complexity after billions of years, deliberate design and building of nanoscale mechanisms requires a means other than reproduction/extinction to winnow successes from failures in proceeding from simplicity to complexity.

Such means are difficult to provide for anything other than small assemblages of atoms viewable by an AFM or STM. Advocates agree this is a valid constraint using current technology, but they insist that this is not a fundamental constraint imposed by the laws of physics. They assert that, once mechanosynthetic tooltips and similar future positionally-controlled molecular tools are fabricated, the same technology could permit prototyping, testing, and rework of failed designs. However, both critics and advocates agree that this expectation remains to be proven and further research will be required to resolve the issue.

In the latest report A Matter of Size: Triennial Review of the National Nanotechnology Initiative put out by the National Academies Press in December 2006, no clear way forward toward molecular nanotechnology could yet be seen, as per the conclusion on page 108 of that report: "Although theoretical calculations can be made today, the eventually attainable range of chemical reaction cycles, error rates, speed of operation, and thermodynamic efficiencies of such bottom-up manufacturing systems cannot be reliably predicted at this time. Thus, the eventually attainable perfection and complexity of manufactured products, while they can be calculated in theory, cannot be predicted with confidence.

Finally, the optimum research paths that might lead to systems which greatly exceed the thermodynamic efficiencies and other capabilities of biological systems cannot be reliably predicted at this time. Research funding that is based on the ability of investigators to produce experimental demonstrations that link to abstract models and guide long-term vision is most appropriate to achieve this goal." This call for research leading to demonstrations is welcomed by groups such as the Nanofactory Collaboration who are specifically seeking experimental successes in diamond mechanosynthesis. Perhaps the eventual "Technology Roadmap for Productive Nanosystems" will offer additional constructive insights.

It is perhaps interesting to ask whether or not most structures consistent with physical law can in fact be manufactured. Such a question is a great deal more difficult to answer than, for example, the four-colour map theorem which was proposed in 1852 and proven in 1976, and it is conceptually impossible to prove the negative of this question since no proof by counter-example can be provided.

Advocates assert that to achieve most of the vision of molecular manufacturing it is not necessary to be able to build "any structure that is compatible with natural law." Rather, it is necessary to be able to build only a sufficient (possibly modest) subset of such structures -- as is true, in fact, of any practical manufacturing process used in the world today, and is true even in biology. In any event, as Richard Feynman once said, "It is scientific only to say what's more likely or less likely, and not to be proving all the time what's possible or impossible."

Diamond mechanosynthesis

There is a growing body of peer-reviewed theoretical work on synthesizing diamond by mechanically removing/adding hydrogen atoms and depositing carbon atoms (a process known as mechanosynthesis). This work is slowly permeating the broader nanoscience community and is being critiqued. For instance, Peng et al. reports that the most-studied mechanosynthesis tooltip motif (DCB6Ge) successfully places a C_2 carbon dimer on a C(110) diamond surface at both 300K (room temperature) and 80K (liquid nitrogen temperature), and that the silicon variant (DCB6Si) also works at 80K but not at 300K. Over 100,000 CPU hours were invested in this latest study.

The DCB6 tooltip motif, initially described by Merkle and Freitas at a Foresight Conference in 2002, was the first complete tooltip ever proposed for diamond mechanosynthesis and remains the only tooltip motif that has been successfully simulated for its intended function on a full 200-atom diamond surface. The tooltips modelled in this work are intended to be used only in carefully controlled environments. Maximum acceptable limits for tooltip translational and rotational misplacement errors are reported in Peng et al.—tooltips must be positioned with great accuracy to avoid bonding the dimer incorrectly.

A working nanofactory would require a variety of well-designed tips for different reactions, and detailed analyses of placing atoms on more complicated surfaces. Although this appears a challenging problem given current resources, many tools will be available to help future researchers: Moore's Law predicts further increases in computer power, semiconductor fabrication techniques continue to approach the nanoscale, and researchers grow ever more skilled at using proteins, ribosomes and DNA to perform novel chemistry.

Tools and techniques

The first observations and size measurements of nano-particles were made during the first decade of the 20th century. They are mostly associated with the name of Zsigmondy who made detailed studies of gold sols and other nanomaterials with sizes down to 10 nm and less. He published a book in 1914. He used ultramicroscope that employs a dark field method for seeing particles with sizes much less than light wavelength.

There are traditional techniques developed during 20th century in Interface and Colloid Science for characterising nanomaterials. These methods include several different techniques for characterising particle size distribution. This characterisation is imperative because many materials that are expected to be nano-sized are actually aggregated in solutions. Some of methods are based on light scattering. Other apply ultrasound, such as ultrasound attenuation spectroscopy for testing concentrated nano-dispersions and microemulsions.

There is also a group of traditional techniques for characterising surface charge or zeta potential of nano-particles in solutions. This information is required for proper system stabilsation, preventing its aggregation or flocculation. These methods include microelectrophoresis, electrophoretic light scattering and electroacoustics. The last one, for instance colloid vibration current method is suitable for characterising concentrated systems.

Next group of nanotechnological techniques include those used for fabrication of nanowires, those used in semiconductor fabrication such as deep ultraviolet lithography, electron beam lithography, focused ion beam machining, nanoimprint lithography, atomic layer deposition, and molecular vapour deposition, and further including molecular self-assembly techniques such as those employing di-block copolymers. However, all of these techniques preceded the nanotech era, and are extensions in the development of scientific advancements rather than techniques which were devised with the sole purpose of creating nanotechnology and which were results of nanotechnology research.

There are several important modern developments. The atomic force microscope (AFM) and the Scanning Tunnelling Microscope (STM) are two early versions of scanning probes that launched nanotechnology. There are other types of scanning probe microscopy, all flowing from the ideas of the scanning confocal microscope developed by Marvin Minsky in 1961 and the scanning acoustic microscope (SAM) developed by Calvin Quate and coworkers in the 1970s, that made it possible to see structures at the nanoscale.

The tip of a scanning probe can also be used to manipulate nanostructures. Feature-oriented scanning-positioning methodology suggested by Rostislav Lapshin appears to be a promising way to implement these nanomanipulations in automatic mode. However, this is still a slow process because of low scanning velocity of the microscope. Various techniques of nanolithography such as dip pen nanolithography, electron beam lithography or nanoimprint lithography were also developed. Lithography is a top-down fabrication technique where a bulk material is reduced in size to nanoscale pattern.

The top-down approach anticipates nanodevices that must be built piece by piece in stages, much as manufactured items are made. Scanning probe microscopy is an important technique both for characterisation and synthesis of nanomaterials. Atomic force microscopes and scanning tunnelling microscopes can be used to look at surfaces and to move atoms around.

By designing different tips for these microscopes, they can be used for carving out structures on surfaces and to help guide self-assembling structures. By using, for example, feature-oriented scanning-positioning approach, atoms can be moved around on a surface with scanning probe microscopy techniques. At present, it is expensive and time-consuming for mass production but very suitable for laboratory experimentation.

In contrast, bottom-up techniques build or grow larger structures atom by atom or molecule by molecule. These techniques include chemical synthesis, self-assembly and positional assembly. Another variation of the bottom-up approach is molecular beam epitaxy or MBE. Researchers at Bell Telephone Laboratories like John R. Arthur. Alfred Y. Cho, and Art C. Gossard developed and implemented MBE as a research tool in the late 1960s and 1970s.

Samples made by MBE were key to the discovery of the fractional quantum Hall effect for which the 1998 Nobel Prize in Physics was awarded. MBE allows scientists to lay down atomically-precise layers of atoms and, in the process, build up complex structures. Important for research on semiconductors, MBE is also widely used to make samples and devices for the newly emerging field of spintronics.

Newer techniques such as Dual Polarisation Interferometry are enabling scientists to measure quantitatively the molecular interactions that take place at the nano-scale. However, new therapeutic products, based on responsive nanomaterials, such as the ultradeformable, stress-sensitive Transfersome vesicles, are under development and already approved for human use in some countries.

Nanotechnology Applications

As of April 24, 2008 The Project on Emerging Nanotechnologies claims that over 609 nanotech products exist, with new ones hitting the market at a pace of 3-4 per week. The project lists all of the products in a database. Most applications are limited to the use of "first generation" passive nanomaterials which includes titanium dioxide in sunscreen, cosmetics and some food products; Carbon allotropes used to produce gecko tape; silver in food packaging, clothing, disinfectants and household appliances; zinc oxide in sunscreens and cosmetics, surface coatings, paints and outdoor furniture varnishes; and cerium oxide as a fuel catalyst.

The National Science Foundation funded researcher David Berube to study the field of nanotechnology. His findings are published in the monograph "Nano-Hype: The Truth Behind the Nanotechnology Buzz". This published study concludes that much of what is sold as "nanotechnology" is in fact a recasting of straightforward materials science, which is leading to a "nanotech industry built solely on selling nanotubes, nanowires, and the like" which will "end up with a few suppliers selling low margin products in huge volumes."

Further applications which require actual manipulation or arrangement of nanoscale components await further research. Though technologies branded with the term 'nano' are sometimes little related to and fall far short of the most ambitious and transformative technological goals of the sort in molecular manufacturing proposals, the term still connotes such ideas. Thus there may be a danger that a "nano bubble" will form, or is forming already, from the use of the term by scientists and entrepreneurs to garner funding, regardless of interest in the transformative possibilities of more ambitious and far-sighted work.

Implications of Nanotechnology

Due to the far-ranging claims that have been made about potential applications of nanotechnology, a number of serious concerns have been raised about what effects these will have on the society if realised, and what action if any is appropriate to mitigate these risks. There are possible dangers that arise with the development of nanotechnology. The Center for Responsible Nanotechnology suggests that new developments could result, among other things, in untraceable weapons of mass destruction, networked cameras for use by the government, and weapons developments fast enough to destabilise arms races.

One area of concern is the effect that industrial-scale manufacturing and use of nanomaterials would have on human health and the environment, as suggested by nanotoxicology research. Groups such as the Center for Responsible Nanotechnology have advocated that nanotechnology should be specially regulated by governments for these reasons. Others counter that overregulation would stifle scientific research and the development of innovations which could greatly benefit mankind.

Other experts, including director of the Woodrow Wilson Center's Project on Emerging Nanotechnologies David Rejeski, have testified that successful commercialisation depends on adequate oversight, risk research strategy, and public engagement. More recently local municipalities have passed or are considering-ordinances requiring nanomaterial manufacturers to disclose the known risks of their products.

The National Institute for Occupational Safety and Health is conducting research on how nanoparticles interact with the body's systems and how workers might be exposed to nano-sized particles in the manufacturing or industrial use of nanomaterials. NIOSH offers interim guidelines for working with nanomaterials consistent with the best scientific knowledge.

Longer-term concerns center on the implications that new technologies will have for society at large, and whether these could possibly lead to either a post scarcity economy, or alternatively exacerbate the wealth gap between developed and developing nations. The effects of nanotechnology on the society as a whole, on human health and the environment, on trade, on security, on food systems and even on the definition of "human", have not been characterised or politicised.

Health and environmental concerns

Some of the recently developed nanoparticle products may have unintended consequences. Researchers have discovered that silver nanoparticles used in socks to reduce foot odour are being released in the wash with possible negative consequences. Silver nanoparticles, which are bacteriostatic, may then destroy beneficial bacteria which are important for breaking down organic matter in waste treatment plants or farms.

A study at the University of Rochester found that when rats breathed in nanoparticles, the particles settled in the brain and lungs, which led to significant increases in biomarkers for inflammation and stress response. A major study published more recently in Nature nanotechnology suggests some forms of carbon nanotubes—a poster child for the "nanotechnology revolution"—could be as harmful as asbestos if inhaled in sufficient quantities.

Anthony Seaton of the Institute of Occupational Medicine in Edinburgh, Scotland, who contributed to the article on carbon nanotubes said "We know that some of them probably have the potential to cause mesothelioma. So those sorts of materials need to be handled very carefully." In the absence of specific nano-regulation forthcoming from governments, Paull and Lyons have called for an exclusion of engineered nanoparticles from organic food.

Regulation of nanotechnology

Calls for tighter regulation of nanotechnology have occurred alongside a growing debate related to the human health and safety risks associated with nanotechnology. Further, there is significant debate about who is responsible for the regulation of nanotechnology. While some non-nanotechnology specific regulatory agencies currently cover some products and processes (to varying degrees)—by "bolting on" nanotechnology to existing regulations—there are clear gaps in these regimes.

Stakeholders concerned by the lack of a regulatory framework to assess and control risks associated with the release of nanoparticles and nanotubes have drawn parallels with bovine spongiform encephalopathy ('mad cow's disease), thalidomide, genetically modified food, nuclear energy, reproductive technologies, biotechnology, and asbestosis.

The Woodrow Wilson Centre's Project on Emerging Technologies conclude (among others) that there is insufficient funding for human health and safety research, and as a result there is currently limited understanding of the human health and safety risks associated with nanotechnology.

The Royal Society report identified a risk of nanoparticles or nanotubes being released during disposal, destruction and recycling, and recommended that "manufacturers of products that fall under extended producer responsibility regimes such as end-of-life regulations publish procedures outlining how these materials will be managed to minimize possible human and environmental exposure".

Reflecting the challenges for ensuring responsible life cycle regulation, the Institute for Food and Agricultural Standards has proposed standards for nanotechnology research and development should be integrated across consumer, worker and environmental standards. They also propose that NGOs and other citizen groups play a meaningful role in the development of these standards.

REFERENCES

Akhlesh Lakhtakia (ed), *The Handbook of Nanotechnology*, Nanometer Structures: Theory, Modeling, and Simulation, SPIE Press, Bellingham, WA, USA., 2004.

Das S, Gates AJ, Abdu HA, Rose GS, Picconatto CA, Ellenbogen JC., "Designs for Ultra-Tiny, Special-Purpose Nanoelectronic Circuits", *IEEE Transactions on Circuits and Systems* I 54 (11): 2528-2540, 2007.

Fritz Allhoff, Patrick Lin, James Moor, and John Weckert (eds.), *Nanoethics: The Ethical and Societal Implications of Nanotechnology,* Hoboken: John Wiley & Sons, 2007.

Ghalanbor Z, Marashi SA, Ranjbar B., "Nanotechnology helps medicine: nanoscale swimmers and their future applications", *Med Hypotheses* 65 (1): 198-199, 2005.

Hari Singh Nalwa, *Encyclopedia of Nanoscience and Nanotechnology* (10-Volume Set), American Scientific Publishers, 2004.

Jumana Boussey, Georges Kamarinos, Laurent Montès (editors), Towards Nanotechnology, "Nano et Micro Technologies", *Hermes Sciences Publ.*, Paris, 2003.

Kubik T, Bogunia-Kubik K, Sugisaka M., "Nanotechnology on duty in medical applications", *Curr Pharm Biotechnol*, 6 (1): 17-33, PMID 15727553, 2005.

Levins CG, Schafmeister CE., "The synthesis of curved and linear structures from a minimal set of monomers", *Journal of Organic Chemistry*, 70, p. 9002, 2005.

Lynn E., *Foster: Nanotechnology: Science, Innovation, and Opportunity*, Prentice Hall, December 21 2005.

Michael Rieth and Wolfram Schommers, *Handbook of Theoretical and Computational Nanotechnology* (10-Volume Set), American Scientific Publishers, 2006.

Narayan RJ, Kumta PN, Sfeir C, Lee D-H, Olton D, Choi D., "Nanostructured Ceramics in Medical Devices: Applications and Prospects", JOM 56 (10): 38-43, 2004.

Shetty RC., "Potential pitfalls of nanotechnology in its applications to medicine: immune incompatibility of nanodevices", *Med Hypotheses* 65 (5): 998-9, 2005.

Taniguchi, N., "On the Basic Concept of 'Nano-Technology'," Proc. Intl. Conf. Prod. London, Part II, *British Society of Precision Engineering*, 1974.

William Sims Bainbridge, *Nanoconvergence: The Unity of Nanoscience, Biotechnology, Information Technology and Cognitive Science*, Prentice Hall, June 27 2007.

Zsigmondy, R., "Colloids and the Ultramicroscope", *J. Wiley and Sons*, NY, 1914.

2

Microscopic Techniques in Nanotechnology

Nanotechnology has the potential to create many new materials and devices with wide-ranging applications, such as in medicine, electronics, and energy production. On the other hand, nanotechnology raises many of the same issues as with any introduction of new technology, including concerns about the toxicity and environmental impact of nanomaterials, and their potential effects on global economics, as well as speculation about various doomsday scenarios

Although the full impact of nanotechnology lies far in the future, the field is already having a significant effect on one business: surface science analytical techniques. To view, analyse and perform experiments on the nanoscale, where the world is measured in nanometers, or billionths of a meter, researchers must employ sophisticated microscopy techniques, most of which are less than 50 years old and whose invention and developments gave birth to nanotechnology.

It is hoped that developments in nanotechnology will make possible their construction by some other means, perhaps using biomimetic principles. However, Drexler and other researchers have proposed that advanced nanotechnology, although perhaps initially implemented by biomimetic means, ultimately could be based on mechanical engineering principles, namely, a manufacturing technology based on the mechanical functionality of these components (such as gears, bearings, motors, and structural members) that would enable programmable, positional assembly to atomic specification.

ATOMIC FORCE MICROSCOPE (AFM)

The atomic force microscope (AFM) or scanning force microscope (SFM) is a very high-resolution type of scanning probe microscope, with demonstrated resolution of fractions of a nanometer, more than 1000 times better than the optical diffraction limit. The precursor to the AFM, the scanning tunnelling microscope, was developed by Gerd Binnig and Heinrich Rohrer in the early 1980s, a development that earned them the Nobel Prize for Physics in 1986. Binnig, Quate and Gerber invented the first AFM in 1986.

The AFM is one of the foremost tools for imaging, measuring and manipulating matter at the nanoscale. The term 'microscope' in the name is actually a misnomer because it implies looking, while in fact the information is gathered by "feeling" the surface with a mechanical probe. Piezoelectric elements that facilitate tiny but accurate and precise movements on (electronic) command enable the very precise scanning.

Principle

The AFM consists of a microscale cantilever with a sharp tip (probe) at its end that is used to scan the specimen surface. The cantilever is typically silicon or silicon nitride with a tip radius of curvature on the order of nanometers. When the tip is brought into proximity of a sample surface, forces between the tip and the sample lead to a deflection of the cantilever according to Hooke's law. Depending on the situation, forces that are measured in AFM include mechanical contact force, Van der Waals forces, capillary forces, chemical bonding, electrostatic forces, magnetic forces, Casimir forces, solvation forces etc. As well as force, additional quantities may simultaneously be measured through the use of specialised types of probe.

Typically, the deflection is measured using a laser spot reflected from the top surface of the cantilever into an array of photodiodes. Other methods that are used include optical interferometry, capacitive sensing or piezoresistive AFM cantilevers. These cantilevers are fabricated with piezoresistive elements that act as a strain gauge. Using a Wheatstone bridge, strain in the AFM cantilever due to deflection can be measured, but this method is not as sensitive as laser deflection or interferometry.

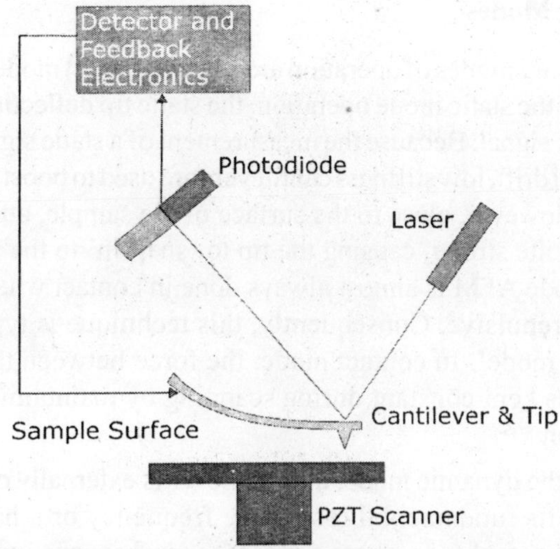

Figure 1. Block Diagram of Atomic Force Microscope

If the tip was scanned at a constant height, a risk would exist that the tip collides with the surface, causing damage. Hence, in most cases a feedback mechanism is employed to adjust the tip-to-sample distance to maintain a constant force between the tip and the sample. Traditionally, the sample is mounted on a piezoelectric tube, that can move the sample in the z direction for maintaining a constant force, and the x and y directions for scanning the sample.

Alternatively a 'tripod' configuration of three piezo crystals may be employed, with each responsible for scanning in the x,y and z directions. This eliminates some of the distortion effects seen with a tube scanner. In newer designs, the tip is mounted on a vertical piezo scanner while the sample is being scanned in X and Y using another piezo block. The resulting map of the area $s = f(x,y)$ represents the topography of the sample.

The AFM can be operated in a number of modes, depending on the application. In general, possible imaging modes are divided into static (also called Contact) modes and a variety of dynamic (or non-contact) modes where the cantilever is vibrated.

Imaging Modes

The primary modes of operation are static (contact) mode and dynamic mode. In the static mode operation, the static tip deflection is used as a feedback signal. Because the measurement of a static signal is prone to noise and drift, low stiffness cantilevers are used to boost the deflection signal. However, close to the surface of the sample, attractive forces can be quite strong, causing the tip to 'snap-in' to the surface. Thus static mode AFM is almost always done in contact where the overall force is repulsive. Consequently, this technique is typically called 'contact mode'. In contact mode, the force between the tip and the surface is kept constant during scanning by maintaining a constant deflection.

In the dynamic mode, the cantilever is externally oscillated at or close to its fundamental resonance frequency or a harmonic. The oscillation amplitude, phase and resonance frequency are modified by tip-sample interaction forces; these changes in oscillation with respect to the external reference oscillation provide information about the sample's characteristics. Schemes for dynamic mode operation include frequency modulation and the more common amplitude modulation.

In frequency modulation, changes in the oscillation frequency provide information about tip-sample interactions. Frequency can be measured with very high sensitivity and thus the frequency modulation mode allows for the use of very stiff cantilevers. Stiff cantilevers provide stability very close to the surface and, as a result, this technique was the first AFM technique to provide true atomic resolution in ultra-high vacuum conditions.

In amplitude modulation, changes in the oscillation amplitude or phase provide the feedback signal for imaging. In amplitude modulation, changes in the phase of oscillation can be used to discriminate between different types of materials on the surface. Amplitude modulation can be operated either in the non-contact or in the intermittent contact regime. In ambient conditions, most samples develop a liquid meniscus layer.

Because of this, keeping the probe tip close enough to the sample for short-range forces to become detectable while preventing the tip from sticking to the surface presents a major hurdle for the non-contact dynamic mode in ambient conditions. Dynamic contact mode was developed to bypass this problem. In dynamic contact mode, the cantilever is oscillated such that the separation distance between the cantilever tip and the sample surface is modulated. Amplitude modulation has also been used in the non-contact regime to image with atomic resolution by using very stiff cantilevers and small amplitudes in an ultra-high vacuum environment.

Tapping Mode

In tapping mode the cantilever is driven to oscillate up and down at near its resonance frequency by a small piezoelectric element mounted in the AFM tip holder. The amplitude of this oscillation is greater than 10 nm, typically 100 to 200 nm. Due to the interaction of forces acting on the cantilever when the tip comes close to the surface, Van der Waals force or dipole-dipole interaction, electrostatic forces, etc cause the amplitude of this oscillation to decrease as the tip gets closer to the sample. An electronic servo uses the piezoelectric actuator to control the height of the cantilever above the sample.

The servo adjusts the height to maintain a set cantilever oscillation amplitude as the cantilever is scanned over the sample. A Tapping AFM image is therefore produced by imaging the force of the oscillating contacts of the tip with the sample surface. This is an improvement on conventional contact AFM, in which the cantilever just drags across the surface at constant force and can result in surface damage. Tapping mode is gentle enough even for the visualisation of supported lipid bilayers or adsorbed single polymer molecules under liquid medium. At the application of proper scanning parameters, the conformation of single molecules remains unchanged for hours.

Force Spectroscopy

Another major application of AFM (besides imaging) is force-spectroscopy, the measurement of force-distance curves. For this method, the AFM tip is extended towards and retracted from the surface as the static deflection of the cantilever is monitored as a function of piezoelectric displacement. These measurements have been used to measure nanoscale contacts, atomic bonding, Van der Waals forces, and Casimir forces, dissolution forces in liquids and single molecule stretching and rupture forces. Forces of the order of a few pico-Newton can now be routinely measured with a vertical distance resolution of better than 0.1 nanometer.

Problems with the technique include no direct measurement of the tip-sample separation and the common need for low stiffness cantilevers which tend to 'snap' to the surface. The snap-in can be reduced by measuring in liquids or by using stiffer cantilevers, but in the latter case a more sensitive deflection sensor is needed. By applying a small dither to the tip, the stiffness (force gradient) of the bond can be measured as well.

Individual Surface Atoms

The AFM can be used to image and manipulate atoms and structures on a variety of surfaces. The atom at the apex of the tip "senses" individual atoms on the underlying surface when it forms incipient chemical bonds with each atom. Because these chemical interactions subtly alter the tip's vibration frequency, they can be detected and mapped. Physicist Oscar Custance and his team used this principle to distinguish between atoms of silicon, tin and lead on an alloy surface.

The trick is to first measure these forces precisely for each type of atom expected in the sample. The team found that the tip interacted most strongly with silicon atoms, and interacted 23% and 41% less strongly with tin and lead atoms. Thus, each different type of atom can be identified in the matrix as the tip is moved across the surface. Such a technique has been used now in biology and extended recently to cell biology. Forces corresponding to

(i) the unbinding of receptor ligand couples

(ii) unfolding of proteins

(iii) cell adhesion at single cell scale have been gathered.

Advantages and Disadvantages

The AFM has several advantages over the scanning electron microscope (SEM). Unlike the electron microscope which provides a two-dimensional projection or a two-dimensional image of a sample, the AFM provides a true three-dimensional surface profile. Additionally, samples viewed by AFM do not require any special treatments (such as metal/carbon coatings) that would irreversibly change or damage the sample.

Figure 2. The first Atomic Force Microscope

While an electron microscope needs an expensive vacuum environment for proper operation, most AFM modes can work perfectly well in ambient air or even a liquid environment. This makes it possible to study biological macromolecules and even living organisms. In principle, AFM can provide higher resolution than SEM. It has been shown to give true atomic resolution in ultra-high vacuum (UHV) and, more recently, in liquid environments. High resolution AFM is comparable in resolution to Scanning Tunnelling Microscopy and Transmission Electron Microscopy.

A disadvantage of AFM compared with the scanning electron microscope (SEM) is the image size. The SEM can image an area on the order of millimetres by millimetres with a depth of field on the order of millimetres. The AFM can only image a maximum height on the order of micrometers and a maximum scanning area of around 150 by 150 micrometers.

Another inconvenience is that an incorrect choice of tip for the required resolution can lead to image artifacts. Traditionally the AFM could not scan images as fast as an SEM, requiring several minutes for a typical scan, while a SEM is capable of scanning at near real-time after the chamber is evacuated.

The relatively slow rate of scanning during AFM imaging often leads to thermal drift in the image, making the AFM microscope less suited for measuring accurate distances between artifacts on the image. However, several fast-acting designs were suggested to increase microscope scanning productivity including what is being termed videoAFM. To eliminate image distortions induced by thermodrift, several methods were also proposed.

AFM images can also be affected by hysteresis of the piezoelectric material and cross-talk between the (x,y,z) axes that may require software enhancement and filtering. Such filtering could "flatten" out real topographical features. However, newer AFM use real-time correction software or closed-loop scanners which practically eliminate these problems. Some AFM also use separated orthogonal scanners (as opposed to a single tube) which also serve to eliminate cross-talk problems. Due to the nature of AFM probes, they cannot normally measure steep walls or overhangs. Specially made cantilevers can be modulated sideways as well as up and down to measure sidewalls, at the cost of more expensive cantilevers and additional artifacts.

Scanning Runneling Microscope (STM)

Scanning tunnelling microscope is a powerful technique for viewing surfaces at the atomic level. Its development in 1981 earned its inventors, Gerd Binnig and Heinrich Rohrer, the Nobel Prize in Physics in 1986. STM probes the density of states of a material using tunnelling current. For STM, good resolution is considered to be 0.1 nm lateral resolution and 0.01 nm depth resolution. The STM can be used not only in ultra high vacuum but also in air and various other liquid or gas ambients, and at temperatures ranging from near 0 kelvin to a few hundred degrees Celsius.

The STM is based on the concept of quantum tunnelling. When a conducting tip is brought very near to a metallic or semiconducting surface, a bias between the two can allow electrons to tunnel through the vacuum between them. For low voltages, this tunnelling current is a function of the local density of states (LDOS) at the Fermi level, Ef, of the sample.

Variations in current as the probe passes over the surface are translated into an image. STM can be a challenging technique, as it requires extremely clean surfaces and sharp tips. First the tip is brought into close proximity of the sample by some coarse sample-to-tip control. The values for common sample-to-tip distance, W, range from about 4-7 Å, which is the equilibrium position between attractive ($3<W<10$Å) and repulsive ($W<3$Å) interactions. Once tunnelling is established, piezoelectric transducers are implemented to move the tip in three directions.

As the tip is rastered across the sample in the x-y plane, the density of states and therefore the tunnel current changes. This change in current with respect to position can be measured itself, or the height, z, of the tip corresponding to a constant current can be measured. These two modes are called constant height mode and constant current mode, respectively. In constant current mode, feedback electronics adjust the height by a voltage to the piezoelectric height control mechanism. This leads to a height variation and thus the image comes from the tip topography across the sample and gives a constant charge density surface; this means contrast on the image is due to variations in charge density.

In constant height, the voltage and height are both held constant while the current changes to keep the voltage from changing; this leads to an image made of current changes over the surface, which can be related to charge density. The benefit to using a constant height mode is that it is faster, as the piezoelectric movements require more time to register the change in constant current mode than the voltage response in constant height mode.

In addition to scanning across the sample, information on the electronic structure of the sample can be obtained by sweeping voltage and measuring current at a specific location. This type of measurement is called scanning tunnelling spectroscopy (STS).

Instrumentation

The components of an STM include scanning tip, piezoelectric controlled height and x,y scanner, coarse sample-to-tip control, vibration isolation system, and computer.

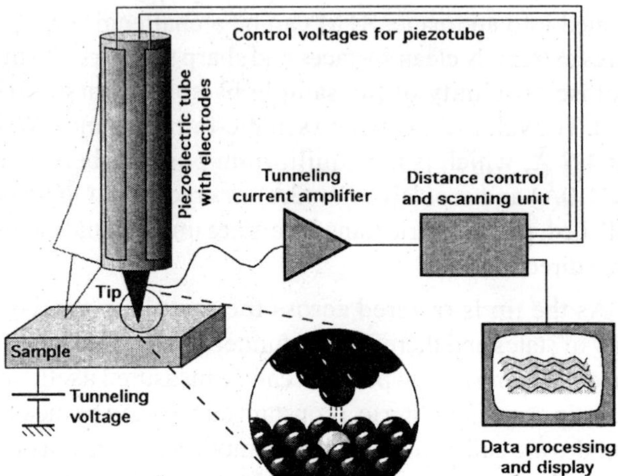

Figure 3. Schematic view of an STM

The resolution of an image is limited by the radius of curvature of the scanning tip of the STM. Additionally, image artifacts can occur if the tip has two tips at the end rather than a single atom; this leads to "double-tip imaging," a situation in which both tips contribute to the tunnelling. Therefore it has been essential to develop processes for consistently obtaining sharp, usable tips. Recently, carbon nanotubes have been used in this instance.

The tip is often made of tungsten or platinum-iridium, though gold is also used. Tungsten tips are usually made by electrochemical etching, and platinum-iridium tips by mechanical shearing. Due to the extreme sensitivity of tunnel current to height, proper vibration isolation is imperative for obtaining usable results. In the first STM by Binnig and Rohrer, magnetic levitation was used to keep the STM free from vibrations; now spring systems are often used.

Additionally, mechanisms for reducing eddy currents are implemented. Maintaining the tip position with respect to the sample, scanning the sample in raster fashion and acquiring the data is computer controlled. The computer is also used for enhancing the image with the help of image processing as well as performing quantitative morphological measurements.

Many other microscopy techniques have been developed based upon STM. These include Photon Scanning Microscopy (PSTM), which uses an optical tip to tunnel photons; Scanning Tunnelling Potentiometry (STP), which measures electric potential across a surface; and spin polarised scanning tunnelling microscopy (SPSTM), which uses a ferromagnetic tip to tunnel spin-polarised electrons into a magnetic sample.

Other STM methods involve manipulating the tip in order to change the topography of the sample. This is attractive for several reasons. Firstly the STM has an atomically precise positioning system which allows very accurate atomic scale manipulation. Furthermore, after the surface is modified by the tip, it is a simple matter to then image with the same tip, without changing the instrument.

IBM researchers developed a way to manipulate Xenon atoms absorbed on a nickel surface This technique has been used to create electron "corrals" with a small number of adsorbed atoms, which allows the STM to be used to observe electron Friedel Oscillations on the surface of the material. Aside from modifying the actual sample surface, one can also use the STM to tunnel electrons into a layer of E-Beam photoresist on a sample, in order to do lithography. This has the advantage of offering more control of the exposure than traditional Electron beam lithography.

Scanning Probe Microscopy (SPM)

Scanning probe microscopy is a branch of microscopy that forms images of surfaces using a physical probe that scans the specimen. An image of the surface is obtained by mechanically moving the probe in a raster scan of the specimen, line by line, and recording the probe-surface interaction as a function of position. SPM was founded with the invention of the scanning tunnelling microscope in 1981.

Many scanning probe microscopes can image several interactions simultaneously. The manner of using these interactions to obtain an image is generally called a mode. The resolution varies somewhat from technique to technique, but some probe techniques reach a rather impressive atomic resolution. They owe this largely to the ability of piezoelectric actuators to execute motions with a precision and accuracy at the atomic level or better on electronic command. One could rightly call this family of technique 'piezoelectric techniques'. The other common denominator is that the data are typically obtained as a two-dimensional grid of data points, visualised in false colour as a computer image.

Advantages

— The resolution of the microscopes is not limited by diffraction, but only by the size of the probe-sample interaction volume, which can be as small as a few picometres. Hence the ability to measure small local differences in object height (like that of 135 picometre steps on <100> silicon) is unparalleled. Laterally the probe-sample interaction extends only across the tip atom or atoms involved in the interaction.

— The interaction can be used to modify the sample to create small structures (nanolithography).

— Unlike electron microscope methods, specimens do not require a partial vacuum but can be observed in air at standard temperature and pressure or while submerged in a liquid reaction vessel.

Disadvantages

— The detailed shape of the scanning tip is sometimes difficult to determine. Its effect on the resulting data is particularly noticeable if the specimen varies greatly in height over lateral distances of 10 nm or less.

— The scanning techniques are generally slower in acquiring images, due to the scanning process. As a result, efforts are being made to greatly improve the scanning rate. Like all scanning techniques, the embedding of spatial information into a time sequence opens the door to uncertainties in metrology, say of lateral spacings and angles, which arise due to time-domain effects like specimen drift, feedback loop oscillation, and mechanical vibration.

— The maximum image size is generally smaller.

— Scanning probe microscopy is often not useful for examining buried solid-solid or liquid-liquid interfaces.

Confocal Microscopy

Confocal microscopy is an optical imaging technique used to increase micrograph contrast and/or to reconstruct three-dimensional images by using a spatial pinhole to eliminate out-of-focus light or flare in specimens that are thicker than the focal plane. This technique has been gaining popularity in the scientific and industrial communities. Typical applications include life sciences and semiconductor inspection.

Concept

The principle of confocal imaging was patented by Marvin Minsky in 1957. In a conventional fluorescence microscope, the entire specimen is flooded in light from a light source. Due to the conservation of light intensity transportation, all parts of the specimen throughout the optical path will be excited and the fluorescence detected by a photodetector or a camera. In contrast, a confocal microscope uses point illumination and a pinhole in an optically conjugate plane in front of the detector to eliminate out-of-focus information.

Only the light within the focal plane can be detected, so the image quality is much better than that of wide-field images. As only one point is illuminated at a time in confocal microscopy, 2D or 3D imaging requires scanning over a regular raster in the specimen. The thickness of the focal plane is defined mostly by the square of the numerical aperture of the objective lens, and also by the optical properties of the specimen and the ambient index of refraction.

Kinds of Confocal Microscopes

Three types of confocal microscopes are commercially available: Confocal laser scanning microscopes, spinning-disk confocal microscopes and Programmable Array Microscopes (PAM). Confocal laser scanning microscopy yields better image quality than Nipkow and PAM, but the imaging frame rate was very slow (less than 3 frames/

second) until recently; spinning-disk confocal microscopes can achieve video rate imaging-a desirable feature for dynamic observations such as live cell imaging. Confocal laser scanning microscopy has now been improved to provide better than video rate (60 frames/second) imaging by using MEMS based scanning mirrors.

Scanning Acoustic Microscope (SAM)

A Scanning Acoustic Microscope is a device which uses focused sound to investigate, measure, or image an object. It is commonly used in failure analysis and non-destructive evaluation. It also has applications in biological and medical research. The semiconductor industry has found the Scanning Acoustic Microscope useful in detecting voids, cracks, and delaminations within microelectronic packages.

Principles

Scanning Acoustic Microscopy works by directing focused sound from a transducer at a small point on a target object. Sound hitting the object is either scattered, absorbed, reflected (scattered at 180°) or transmitted (scattered at 0°). Typically, either the reflected or transmitted sound is gathered and measured. Based on the measurement, a value is assigned to the location investigated. The transducer (or object) is moved slightly and then insonified again. This process is repeated in a systematic pattern until the entire region of interest has been investigated.

Often the values for each point are assembled into an image of the object. The contrast seen in the image is based either on the object's geometry or material composition. The resolution of the image is limited either by the physical scanning resolution or the width of the sound beam (which in turn is determined by the frequency of the sound).

Applications of Scanning Acoustic Microscopy

Medicine and biology

SAM can provide data on the elasticity of cells and tissues, which can give useful information on the physical forces holding structures in a particular shape and the mechanics of structures such as the cytoskeleton. These studies are particularly valuable in investigating processes such as cell motility.

— Flaw Detection
— Internal stress investigation
— Elastic property characterisation

Ultrasonic Direct Microscopy (UDM)

Ultrasonic Direct Microscope base on measuring the ultrasonic signal directly. The conventional sensor include a level sensor that means that the measurement instrument attitude of the signal. ADM technique deal with ultrasonic signal as is. That make possible to produce modulate ultrasonic signal and to check not only the main frequency, and enable wide range in order to examine little chances in to material in test.

Ultramicroscope

The ultramicroscope is a system of illumination for extremely small objects such as colloidal particles, fog droplets, or smoke particles. The objects are held in liquid or gaseous suspension in an enclosure with an intensely black background (usually a black body) and illuminated with a convergent pencil of very bright light entering from one side and coming to focus in the field of view—the "Tyndall cone" familiar in experiements on scattering.

With this arrangement, objects too small to form visible images in the microscope produce small diffraction ring patterns that appear as bright specks on a dark field. Ultramicroscopes are used in the study of Brownian motion, in the Millikan droplet experiment for measuring the electric charge of the electron, and in observing ionization tracks in cloud chambers.

References

Hari Singh Nalwa . *Encyclopedia of Nanoscience and Nanotechnology* (10-Volume Set), American Scientific Publishers. 2004.

Michael Rieth and Wolfram Schommers. *Handbook of Theoretical and Computational Nanotechnology*. American Scientific Publishers. 2006.

Akhlesh Lakhtakia (ed) *The Handbook of Nanotechnology. Nanometer Structures: Theory, Modeling, and Simulation*. SPIE Press, Bellingham, WA, USA. 2004.

Fei Wang & Akhlesh Lakhtakia (eds). *Selected Papers on Nanotechnology — Theory & Modeling (Milestone Volume 182)*. SPIE Press, Bellingham, WA, USA.2006

3

Nanorobotics

The engineering of molecular products needs to be carried out by robotic devices, which have been termed as nanorobots. A nanorobot is essentially a controllable machine at the nanometer or molecular scale that is composed of nanoscale components. The field of nanorobotics studies the design, manufacturing, programming, and control of the nanoscale robots.

Nanorobots would constitute any passive or active structure capable of actuation, sensing, signalling, information processing, intelligence, swarm behaviour at the nano scale. These functionalities could be illustrated individually by a nanorobot or in combinations of nanorobots. Some of the abilities that are desirable for a nanorobot to function are:

1. *Swarm Intelligence*-decentralisation and distributive intelligence
2. *Cooperative behaviour*-emergent and evolutionary behaviour
3. *Self-assembly and replication*-assemblage at nano scale and "nanomaintenance"
4. *Nanoinformation processing and programmability* -for programming and controlling nanorobots
5. *Nano to macroworld interface architecture* - an architecture enabling instant access to the nanorobots and its control and maintenance

There are many differences between macro and nanoscale robots. These occur mainly in the basic laws that govern their dynamics. Macroscaled robots are essentially in the Newtonian mechanics domain whereas the laws governing nanorobots are in the molecular quantum mechanics domain. Furthermore, uncertainty plays a crucial role in nanorobotic systems. The fundamental barrier for dealing with uncertainty at the nano scale is imposed by the quantum and the statistical mechanics and thermal excitations.

For a certain nanosystem at some particular temperature, there are positional uncertainties, which cannot be modified or further reduced. The nanorobots are invisible to naked eye, which makes them hard to manipulate and work with. Techniques like scanning electron microscopy (SEM) and atomic force microscopy (AFM) are being employed to establish a visual and haptic interface to enable us to sense the molecular structure of these nanoscaled devices. Virtual reality (VR) techniques are currently being explored in nanoscience and biotechnology research as a way to enhance the operator's perception (vision and haptics) by approaching more or less a state of "full immersion" or "telepresence."

The development of nanorobots or nanomachine components presents difficult fabrication and control challenges. Such devices will operate in microenvironments whose physical properties differ from those encountered by conventional parts. Since these nanoscale devices have not yet been fabricated, evaluating possible designs and control algorithms requires using theoretical estimates and virtual interfaces/environments. Such interfaces/simulations can operate at various levels of detail to trade-off physical accuracy, computational cost, number of components and the time over which the simulation follows the nanoobject behaviours. They can enable nanoscientists to extend their eyes and hands into the nanoworld and also enable new types of exploration and whole new classes of experiments in the biological and physical sciences.

VR simulations can also be used to develop virtual assemblies of nano and bio-nanocomponents into mobile linkages and predict their performance. Nanorobots with completely artificial components have not been realised yet. The active area of research in this field is focused more on molecular robots, which are thoroughly inspired by nature's way of doing things at nano scale. Mother Nature has her own set of molecular machines that have been working for centuries, and have been optimized for performance and design over the ages. As the knowledge and understanding of these numerous machines continues to increase, we now see a possibility of using the natural machines, or creating synthetic ones from scratch, using nature's components.

The main goal in the field of molecular machines is to use various biological elements-whose function at the cellular level creates motion, force or a signal-as machine components. These components perform their preprogrammed biological function in response to the specific physiochemical stimuli but in an artificial setting. In this way proteins and DNA could act as motors, mechanical joints, transmission elements, or sensors. If all these different components were assembled together in the proper proportion and orientation they would form nanodevices with multiple degrees of freedom, able to apply forces and manipulate objects in the nanoscale world.

The advantage of using nature's machine components is that they are highly efficient and reliable. Nanorobotics is a field, which calls for collaborative efforts between physicists, chemists, biologists, computer scientists, engineers, and other specialists to work towards this common objective. Currently this field is still evolving, but several substantial steps have been taken by great researchers all over the world and are contributing to this ever challenging and exciting field.

The ability to manipulate matter at the nano scale is one core application for which nanorobots could be the technological solution. A lot has been written in the literature about the significance and motivation behind constructing a nanorobot. The applications range from medical to environmental sensing to space and military applications. Molecular construction of complex devices could be possible by nanorobots of the future. From precise drug delivery to repairing cells and fighting tumour cells, nanorobots are expected to revolutionise the medical industry in the future. These applications come under the field of nanomedicine, which is a very active area of research in nanotechnology. These molecular machines hence form the basic enablers of future nanoscale applications.

Natural Molecular Machines

The molecular machines into three broad categories - protein-based, DNA-based, and chemical molecular motors.

Protein-Based Molecular Machines

1. ATP Synthase
2. The kinesin, myosin, dynein, and agella molecular motors

ATP synthase

Synthesis of ATP is carried out by an enzyme, known as ATP Synthase. The inner mitochondrial membrane contains the ATP Synthase. The ATP Synthase is actually a combination of two motors functioning together as described in Figure 1.

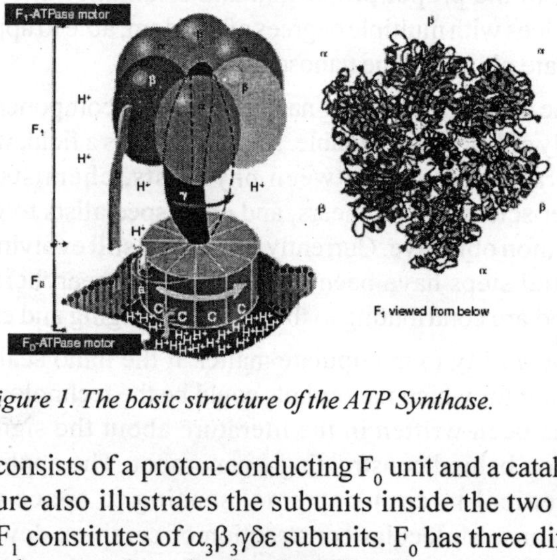

Figure 1. The basic structure of the ATP Synthase.

This enzyme consists of a proton-conducting F_0 unit and a catalytic F_1 unit. The figure also illustrates the subunits inside the two motor components. F_1 constitutes of $\alpha_3\beta_3\gamma\delta\varepsilon$ subunits. F_0 has three different protein molecules, namely, subunit a, b, and c. The γ subunit of F_1 is attached to the c subunit of F_0 and is hence rotated along with it. The $\alpha_3\beta_3$ subunits are fixed to the b subunit of F_0 and hence do not move. Further the b subunit is held inside the membrane by a subunit of F_0.

ATP Synthase "nano" properties

Reversibility of the ATP Synthase: There are two directions in ATP Synthase system and these two directions correspond to two different functionalities and behaviour. This two-way behaviour is because of the reversible nature of the ATP-ADP cycle and the structure of the ATP Synthase. Let us term the forward direction as when the F_0 drives the γ subunit of F_1 and hence ATP synthesis takes place. And the backward direction is when hydrolysis of ATP counter-rotated the γ subunit and hence the F_0 motor and leads to pumping back the protons. Therefore, the forward direction is powered by the proton motive force and the backward direction is powered by the ATP hydrolysis. Which particular direction is being followed depends upon the situation and the environmental factors around the ATP Synthase.

Coupling of Proton Flow (F_0) and the ATP Synthesis and Hydrolysis (in F_1): Boyer proposed a model which predicted that the F_0 and F_1 motors are connected through the γ subunit. Further he proposed that this connection was mechanical in nature. Figure 2 illustrates the electrostatic surface potential on α3β3 συβυνιτσ.

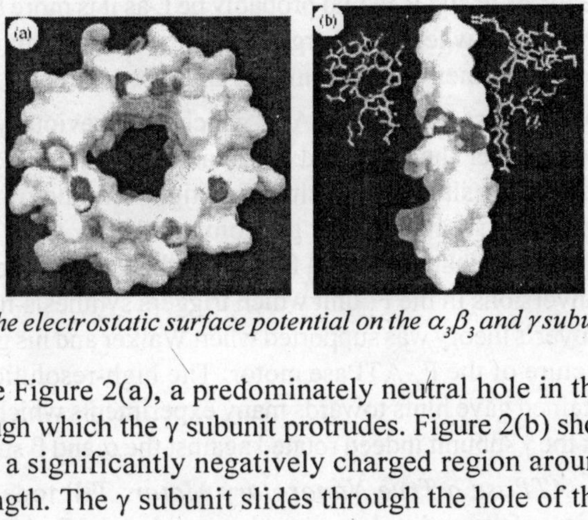

Figure 2. The electrostatic surface potential on the $\alpha_3\beta_3$ and γsubunits.

Shown in the Figure 2(a), a predominately neutral hole in the $\alpha_3\beta_3$ subunit through which the γ subunit protrudes. Figure 2(b) shows a γ subunit with a significantly negatively charged region around half across its length. The γ subunit slides through the hole of the $\alpha_3\beta_3$ subunits.

Boyer's Binding Change Mechanism: Boyer isolated the F_1 part of the ATP Synthase complex. It was found that the a and β subunits alternate in this cylindrical part of the F_1 structure. As per this model each a and β pair forms a catalytic site. The rotation of the γ subunit induces structural conformation in the $\alpha_3\beta_3$ subunits. Although the three catalytic units are identical in their chemistry, they are functionally very different at any given point in time. These conformal changes induce a change in the binding affinities of all the three catalytic sites towards the ATPase reactants (ADP, Pi, ATP, etc.). Figure 3 shows the binding change mechanism as proposed by Boyer.

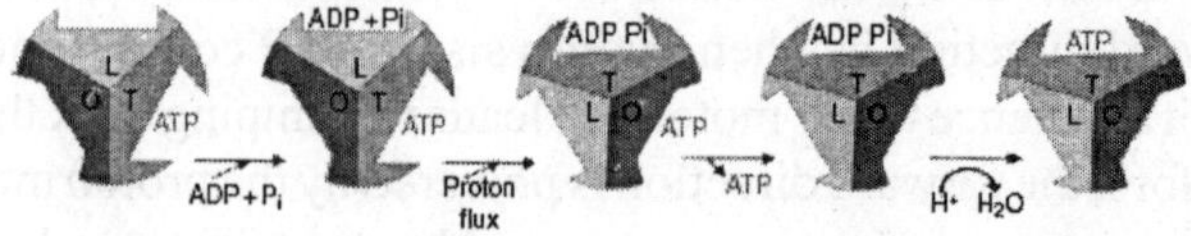

Figure 3. Boyer's binding change mechanism.

The three catalytic sites could be in three distinct forms. O form stands for open which implies that it has got very low affinity for reactants; L form stands for loose, which implies that it would bind the reactants loosely but is still catalytically inactive; T form stands for tight, which binds the reactant tightly and is catalytically active. The situation depicted in Figure 3 shows one of the sites being bound to ATP at T. The site for ADP and P_i would probably be L as it is more binding than the O site. Now when the energy is input to the system the sites are changed. T becomes O, O becomes L, and L becomes T.

Due to these changes the ATP which was previously in T is now released. Further the ADP and P_i are now tightly bound due to the conversion of L site to the T site. This tight binding of ADP and P_i allows them to spontaneously get converted into ATP. Hence, this model proposed that the proton flow from the F_0 is coupled to the site inter conversions in the F_1 unit which triggers synthesis-hydrolysis of ATP. Boyer's theory was supported when Walker and his group solved the structure of the F_1-ATPase motor. The high-resolution structure thus obtained gave hints towards many experiments which proved the fact that the γ subunit indeed rotated against the α and β subunits.

F_1-ATPase a True Nanorotary Motor: Till today the exact mechanism of the molecular motor characterised by F_1-ATpase has not been fully determined. Research by Kinosita's lab is a step towards this goal and proposes some very conclusive models for the same. The results obtained show not only the various methods through which we can analyse these nanodevices, but also predicts many characteristics for these.

What is known till now is that the γ subunit rotates inside the α-β hexamer, but whether the rotation is continuous or random was not known. Kinosita's lab solved this problem by imaging the F_1-ATPase molecule. The objective of their experiment was to determine the uniqueness of the rotary motion and its characteristics. Figure 4 depicts the experiment that was performed by this group. They attached a micrometer long actin filament to the γ subunit. This actin filament was fluorescently labelled, so that its fluorescence could be measured under a microscope.

Figure 4. Experiment performed for Imaging of F1-ATPase.

Hydrolysis of the ATP (when introduced in the experiment) led to the rotation of the γ subunit and in effect the rotation of the actin filament. Not all the actin filaments were observed to have rotation. But some percentage of them did rotate and that too in a unique direction and without having much reversibility in the direction. This direct imaging proved that the structure solved by Walker and group was indeed correct and there exists rotary motion between γ subunit and the α and β hexamer.

The rotation is not restricted to some particular rotational angle but is continuously progressive in a particular direction. As the rotational motion is a continuous one, possibility of it being a twisting motion is ruled out. Hence, there should be no direct linkage between the γ subunit and the α-β hexamer in F_1-ATPase. It was further observed by the authors that the rotation of the actin filament was in certain steps.

Other observed behaviours of the F_1-ATPase motor

Rotational Rates: Rotational rates were dependent upon the length of the actin filament. Rotational rates were found to be inversely proportional to the filament length. This could be attributed to the fact that the hydrodynamic friction is proportional to the cube of the length of the filament, therefore, longer length implied that the frictional forces were higher and rotational speeds in effect slower. As reported in the principle article, if the filament were to rotate at the speed of 6 rev/sec and the length of the actin filament were 1 μm, then as per the relation:

$$N = \omega\xi$$

The torque required would be 40 pN nm. Here ξ is the drag coefficient given by the relation:

$$\xi = \frac{4\pi}{3}\eta L^3 \frac{1}{\left[1n(L/r) - 0.447\right]}$$

where μ is the viscosity of the medium = 10^{-3} N sec/m^2 and r = 5 nm, the radius of the filament.

Efficiency of F_1 - Work done in one step, that is, 120° rotation is equal to the torque generated times the angular displacement for that step. This comes to about 80 pN nm. The free energy obtained by the hydrolysis of one ATP molecule is:

$$\Delta G = \Delta G_0 + k_B T \frac{In[ADP][Pi]}{[ATP]}$$

Here, the standard free energy change per ATP molecule is equal to -50 pNnm at pH 7. The thermal energy at room temperature is equal to 4.1 pN nm. At intracellular conditions, [ATP] and [Pi] is in the order of 10^{-3} M. Therefore, for [ADP] of 50 μM, ΔG is equal to -90.604 pN nm. Comparing this value of the free energy obtained by the hydrolysis of one molecule of ATP with the mechanical work done by the motor, we can deduce that the F_1-ATPase motor works on efficiency close to 100%!

Chemical Synthesis of ATP Powered by Mechanical Energy: In another study, evidence has been provided to justify the claim that the chemical synthesis of ATP occurs when propelled by Mechanical energy. This basic nature of the F_1-ATPase was known for quite some time, but it is the first time that it is being experimentally verified. To prove this concept, the F_1 was bound to the glass surface through histidine residues attached at the end of the γ subunits. A magnetic bead coated with streptavidin was attached to the γ subunit (Figure 5).

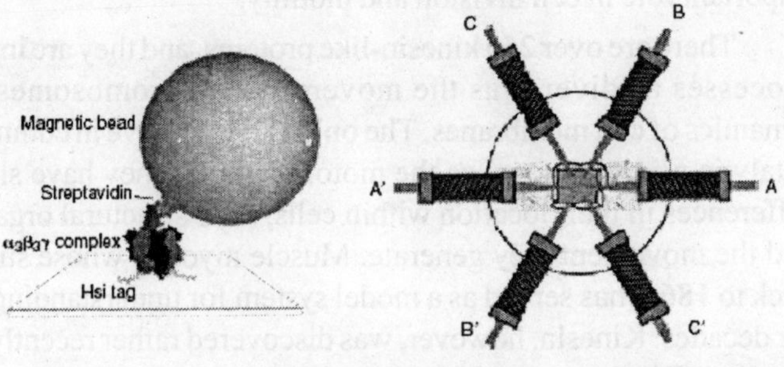

Figure 5. Magnetic bead is attached to the γ subunit here (left figure) and the arrangement of the magnets (right figure).

Electric magnets were used to rotate this bead attached to the γ subunit. The rotation resulted in appearance of ATP in the medium (which was initially immersed in ADP). Thus, the connection between the syntheses of ATP as a result of the mechanical energy input is established. The exact mechanism of the F_1-ATPase rotation is still an active area of research today and many groups are working towards finding it.

The key to solving the mechanism is solving the transient conformation of the catalytic sites and the γ subunit when rotation is taking place. What is not clear is the correspondence between the chemical reactions at the catalytic sites and their influence on the rotation of the γ subunit. Which event triggers the rotation and which does not has still to be exactly determined? Many models have been predicted, but they all still elude the reality of the rotational mechanism.

Kinesin, Myosin, Dynein, and Flagella molecular motors

Molecular motors that move unidirectionally along protein polymers (actin or microtubules) drive the motions of muscles as well as much smaller intracellular cargoes. In addition to the F_0-F_1-ATPase motors inside the cell, there are linear transport motors present as tiny vehicles known as motor proteins that transport molecular cargoes that also require ATP for functioning. These minute cellular machines exist in three families—the kinesins, the myosins, and the dyneins. The cargoes can be organelles, lipids, or proteins etc. They play an important role in cell division and motility.

There are over 250 kinesin-like proteins, and they are involved in processes as diverse as the movement of chromosomes and the dynamics of cell membranes. The only part they have in common is the catalytic portion known as the motor domain. They have significant differences in their location within cells, their structural organisation, and the movement they generate. Muscle myosin, whose study dates back to 1864, has served as a model system for understanding motility for decades. Kinesin, however, was discovered rather recently using in vitro motility assays in 1985.

Conventional kinesin is a highly processive motor that can take several hundred steps on a microtubule without detaching whereas muscle myosin executes a single "stroke" and then dissociates. Kinesin and myosin make up for an interesting comparison. Kinesin is microtubule-based; it binds to and carries cargoes along microtubules whereas myosin is actin-based. The motor domain of kinesin weighs one third the size of that of myosin and one tenth of that of dynein. Before the advent of modern microscopic and analytic techniques, it was believed that these two have little in common. However, the crystal structures available today indicate that they probably originated from a common ancestor.

Myosin linear motor

Myosin is a diverse superfamily of motor proteins. Myosin-based molecular machines transport cargoes along actin filaments—the two stranded helical polymers of the protein actin, about 5 to 9 nm in diameter. They do this by hydrolysing ATP and utilising the energy released. In addition to transport, they are also involved in the process of force generation during muscle contraction, wherein thin actin filaments and thick myosin filaments slide past each other. Not all members of the myosin superfamily have been characterised as of now. However, much is known about the structure and function.

Myosin molecules were first sighted through electron microscope protruding out from thick filaments and interacting with the thin actin filaments in late 1950s. Since then it was known that ATP plays a role in myosin related muscle movement along actin. However, the exact mechanism was unknown, which was explained later in 1971 by Lymn and Taylor.

Structure of Myosin Molecular Motor: Myosin molecule has a size of about 520 kD including two 220 kD heavy chains and light chains of sizes between 15 and 22 kD. They can be visualised as two identical globular "motor heads," also known as motor domains, each comprising of a catalytic domain (actin, nucleotide as well as light chain binding sites) and about 8 nmlong lever arms. The heads, also sometimes referred to as S1 regions (subfragment 1) are shown in blue, while the lever arms or the light chains, in yellow.

Both these heads are connected via a coiled coil made of two a-helical coils (gray) to the thick base filament. The light chains have considerable sequence similarity with the protein "calmodulin" and troponin C, and are sometimes referred to as calmodulin-like chains. They act as links to the motor domains and do not play any role in their ATP binding activity but for some exceptions. The motor domain in itself is sufficient for moving actin filaments. Three-dimensional (3D) structures of myosin head revealed that it is a pear-shaped domain, about 19 nm long and 5 nm in maximum diameter

.Function of Myosin Molecular Motor: A crossbridge-cycle model for the action of myosin on actin has been widely accepted since 1957. Since the atomic structures of actin monomer and myosin were resolved, this model has been refined into a "lever-arm model," which is now acceptable. Only one motor head is able to connect to the actin filament at a time, the other head remains passive. Initially, the catalytic domain in the head has ADP and P_i bound to it and as a result, its binding with actin is weak. With the active motor head docking properly to the actinbinding site, the Pi has to be released.

As soon as this happens, the lever arm swings counter clock wise due to a conformational change. This pushes the actin filament down by about 10 nm along its longitudinal axis. The active motor head now releases its bound ADP and another ATP molecule by way of Brownian motion quickly replaces it, making the binding of the head to the actin filament weak again. The myosin motor then dissociates from the actin filament, and a new cycle starts. However, nanomanipulation of single S1 molecules (motor domains) show that myosin can takemultiple steps per ATP molecule hydrolysed, moving in 5.3 nm steps and resulting in displacements of 11 to 30 nm.

Kinesin linear motor

Kinesin and dynein family of proteins are involved in cellular cargo transport along microtubules as opposed to actin in the case of myosin. Microtubules are 25 nm diameter tubes made of protein tubulin and are present in the cells in an organised manner. Microtubules have polarity; one end being the plus (fast growing) end while the other end is the minus end. Kinesins move from minus end to plus end, while dyneins move from plus end to the minus end of the microtubules. Microtubule arrangement varies in different cell systems. In nerve axons, they are arranged longitudinally in such a manner that their plus ends point away from the cell body and into the axon.

In epithelial cells, their plus end points towards the basement membrane. They deviate radially out of the cell centre in fibroblasts and macrophages with the plus end protruding outwards. Like myosin, kinesin is also an ATP-driven motor. One unique characteristic of kinesin family of proteins is their processivity—they bind to microtubules and literally "walk" on it for many enzymatic cycles before detaching. Also, each of the globular heads/motor domains of kinesin is made of one single polypeptide unlike myosin.

Structure of Kinesin Molecular Motor: A lot of structural information about kinesin is now available through the crystal structures. The motor domain contains a folding motif similar to that of myosin and G proteins. The two heads or the motor domains of kinesin are linked via "neck linkers" to a long coiled coil, which extends up to the cargo. They interact with the a and β subunits of the tubulin hetrodimer along the microtubule proto filament. The heads have the nucleotide and the microtubule binding domains in them.

Function of Kinesin Molecular Motor: While Kinesin is also a two-headed linear motor, its modus operandi is different from myosin in the sense that both its head work together in a coordinated manner rather than one was being left out. Each of the motor heads is near the microtubule in the initial state with each motor head carrying an ADP molecule. When one of the heads loosely binds to the microtubule, it looses its ADP molecule to facilitate a stronger binding.

Another ATP molecule replaces the ADP, which facilitates a conformational change such that the neck region of the bound head snaps forward and zips on to the head. In the process it pulls the other ADP carrying motor head forward by about 16 nm so that it can bind to the next microtubule-binding site. This results in the net movement of the cargo by about 8 nm. The second head now binds to the microtubule by losing its ADP, which is promptly replaced by another ATP molecule due to Brownian motion.

The first head meanwhile hydrolyses the ATP and loses the resulting P_i. It is then snapped forward by the second head while it carries its ADP forward. Hence, coordinated hydrolysis of ATP in the two motor heads is the key to the kinesin processivity. Kinesin is able to take about 100 steps before detaching from the microtubule while moving at 1000 nm/sec and exerting forces of the order of 5 to 6 pN at stall.

Dynein motor

The Dynein superfamily of proteins was introduced in 1965. Dyneins exist in two isoforms, the cytoplasmic and the axonemal. Cytoplasmic dyneins are involved in cargo movement, while axonemal dyneins are involved in producing bending motions of cilia and agella.

Structure of Dynein MolecularMotor: The structure consists of two heavy chains in the form of globular heads, three intermediate chains and four light intermediate chains. Recent studies have exposed a linker domain connecting the "stem" region below the heads to the head itself. Also from the top of the heads the microtubule binding domains (the stalk region, not visible in the figure) protrude out. The ends of these stalks have smaller ATP sensitive globular domains, which bind to the microtubules.

Cytoplasmic dynein is associated with a protein complex known as dynactin, which contains ten subunits. Some of them are shown in the figure as p150, p135, actin-related protein 1 (Arp1), actin, dynamitin, capping protein, and p62 subunit. These play an important regulatory role in the binding ability of dynein to the microtubules. The heavy chains forming the two globular heads contain the ATPase and microtubule motor domains.

One striking difference that dynein exhibits compared to kinesins and myosins is that dynein has AAA modules, which indicate that its mode of working will be entirely different from kinesins and myosins. This puts dyneins into the AAA superfamily of mechanoenzymes. The dynein heavy chains contain six tandemly linked AAA modules with the head having a ring-like domain organisation, typical of AAA superfamily. Four of these are nucleotide binding motifs, named P1-P4, but only P1 (AAA1) is able to hydrolyse ATP.

Function of Dynein MolecularMotor: Because dynein is larger and more complex structure as compared to other motor proteins, its mode of operation is not as well known. However, very recently, Burgess et al. have used electron microscopy and image processing to show the structure of a flagellar dynein at the start and end of its power stroke, giving some insight into its possible mode of force generation. When the dynein contains bound ADP and V_i (vandate), it is in the prepower stroke conformation. The state when it has lost the two, known as the apostate is the more compact postpower stroke state. There is a distinct conformational change involving the stem, linker, head, and the stalk that produces about 15 nm of translation onto the microtubule bound to the stalk.

Flagella motors

Unicellular organisms, such as, Escherichia coli have an interesting mode of motility. They have a number of molecular motors, about 45 nm in diameter, that drive their "feet" or the flagella that help the cell to swim. Motility is critical for cells, as they often have to travel from a less favourable to a more favourable environment. The flagella are helical filaments that extend out of the cell into the medium and perform a function analogous to what the oars perform to a boat. The flagella and the motor assembly are called a flagellum.

The flagella motors impart a rotary motion into the flagella. In addition to a rotary mechanism, the flagella machines consist of components such as rate meters, particle counters, and gearboxes. These are necessary to help the cell decide which way to go, depending on the change of concentration of nutrients in the surroundings. The rotary motion imparted to the flagella needs to be modulated to ensure the cell is moving in the proper direction as well as all flagella of the given cell are providing a concerted effort towards it. When the motors rotate the flagella in a counter clock wise direction as viewed along the flagella filament from outside, the helical.agella create a wave away from the cell body.

Adjacent flagella subsequently intertwine in a propulsive corkscrew manner and propel the bacteria. When the motors rotate clockwise, the flagellay apart, causing the bacteria to tumble, or change its direction. These reversals occur randomly, giving the bacterium a "random walk," unless of course, there is a preferential direction of motility due to reasons mentioned earlier. The flagella motors allow the bacteria to move at speeds of as much as 25 μm/sec with directional reversals occurring approximately 1 per sec.

A number of bacterial species in addition to E. coli, depend on flagella motors for motility. Some of these are Salmonella enterica serovar Typhimurium (Salmonella), Streptococcus, Vibrio spp., Caulobacter, Leptospira, Aquaspirrilum serpens, and Bacillus. The rotation of flagella motors is stimulated by a flow of ions through them which is a result of a build-up of a transmembrane ion gradient. There is no direct ATP involvement, however, the proton gradient needed for the functioning of flagella motors can be produced by ATPase.

Structure of the Flagella Motors: A complete part list of the flagella motors may not be available as of now. Continued efforts dating back to early 1970s have, however, revealed much of their structure, composition, genetics, and function. Newer models of the motor function are still being proposed with an aim to explain observed experimental phenomena. That means that we do not fully understand the functioning of this motor.

A typical flagella motor from E. coli consists of about 20 different proteins, while there are yet more that are involved in the assembly and functioning. There are 14 Flg-type proteins named FlgA to FlgN; 5 Flh-type proteins called FlhA to FlhE; 19 Fli-type proteins named FliA to FliT; MotA and MotB making a total of 40 related proteins. The name groups Flg, Flh, Fli, and Flg originate from the names of the corresponding genes.

Out of these the main structural proteins are FliC or the filament; FliD (.lament cap); FliF or the MS-ring; FliG; FliM, and FliN (C-ring); FlgB, FlgC, and FlgF (proximal rod); FlgG (distal rod); FlgH (L-ring); FlgI (P-ring); FlgK and FlgL (hook-filament junction); and MotA-MotB. Earlier it was believed that the Mand S are two separate rings andMwas named after membrane and S after supramembranous. Now they are jointly called the MS-ring as it has been found that they are two domains of the same protein FliF. The C-ring is named after cytoplasmic, while the names of the P and L-rings come from "peptidoglycan" and "lipopolysaccharide," respectively. The FlhA, B, FliH, I, O, P, Q, R constitute the "transport apparatus."

The hook and filament part of the flagellum is located outside the cell body, while the motor portion is embedded in the cell membrane with parts that are inside the inner membrane in the cytoplasmic region. MotA and MotB are arranged in a circular array embedded in the inner membrane, with the MS-ring at the centre. Connected to the MS-ring is the proximal end of a shaft, to which the P-ring, which is embedded in the peptidoglycan layer, is attached. Moving further outwards, there is the L-ring embedded in the outer cell membrane followed by the distal shaft end that protrudes out of the cell. To this end there is an attachment of the hook and the filament, both of which are polymers of hook-protein and flagellin, respectively.

Function of the Flagella Motors: The flagellar motors in most cases are powered by protons flowing through the cell membrane barring exceptions such as certain marine bacteria, for example, the Vibrio spp., which are driven by Na+ ions. There are about 1200 protons required to rotate the motor by one rotation. A complete explanation of how this proton flow is able to generate torque is not available as of today. From what is known, the stator units of MotA and MotB play an important role in torque generation. They form a MotA/MotB complex which when oriented properly binds to the peptidoglycan and opens proton channels through which protons can flow. It is believed that there are eight such channels per motor.

The protonmotive force is a result of the difference of pH in the outside and the inside of the cell. The E. coli cells like to maintain a pH of 7.6-7.8 on their inside, hence depending on the pH of the surroundings, the protonmotive force will vary, and hence the speed of rotation of their motors. To test how the speed of rotation depends on the protonmotive force, the motors were powered by external voltage with markers acting as heavy loads attached to them. It was found that the rotation indeed depends directly on the protonmotive force.

According to the most widely accepted model, MotA/MotB complex interacts with the rotor via binding sites. The passage of protons through aMotA/MotB complex (stator or torque generator) moves it so that they bind to the next available binding site on the rotor, thereby stretching their linkage. When the linkage recoils, the rotor assembly has to rotate by one step. Hence, whichever complex receives protons from the flux will rotate the rotor, generating torque. The torque-speed dependence of the motor has been studied in detail and indicates the torque range of about 2700 to 4600 pN nm.

DNA-Based Molecular Machines

There was no mechanical work assigned to it. Energy conversion, trafficking, sensing, etc., were the tasks assigned mainly to proteins. Probably for this reason, DNA turns out to be a simpler structure—with only four kinds of nucleotide bases adenosine, thiamine, guanine, and cytosine (A, T, G, and C) attached in a linear fashion that take a double helical conformation when paired with a complementary strand. Such structural simplicity vis-à-vis proteins—made of 20 odd amino acids with complex folding patterns—results in a simpler structure and predictable behaviour. There are certain qualities that make DNA an attractive choice for the construction of artificial nanomachines.

In recent years, DNA has found use in not only mechanochemical, but also in nanoelectronic systems as well. A DNA double-helical molecule is about 2 nm in diameter and has 3.4-3.6nm helical pitch no matter what its base composition is; a structural uniformity is not achievable with protein structures if one changes their sequence. Furthermore, double-stranded DNA (ds-DNA) has a respectable persistence length of about 50 nm which provides it enough rigidity to be a candidate component of molecular machinery. Single stranded DNA (ss-DNA) is very flexible and cannot be used where rigidity is required, however, this flexibility allows its application in machine components like hinges or nanoactuators. Its persistence length is about 1 nm covering up to three base pairs at 1 M salt concentration.

Other than the above structural features there are two important and exclusive properties that make DNA suitable for molecular level constructions. Theseۥ are-molecular recognition and self-assembly. The nucleotide bases A and T on two different ss-DNA have affinity towards each other, so doGand C. Effective and stable ds-DNA structures are only formed if the base orders of the individual strands are complementary.

Hence, if two complementary single strands of DNA are in a solution, they will eventually recognise each other and hybridize or zip-up forming a ds-DNA. This property of molecular recognition and selfassembly has been exploited in a number of ways to build complex molecular structures. In the mechanical perspective, if the free energy released by hybridisation of two complementary DNA strands is used to lift a hypothetical load, a force capacity of 15 pN can be achieved, comparable to that of other molecular machines like kinesin (5 pN).

Dr. Nadrian Seeman and colleagues built the first artificial DNA-based structure in the form of a cube in 1991. They then went on to create more complex structures such as knots and Borromean rings. In addition to these individual constructs, 2D arrays with the help of the double-crossover (DX) DNA molecule. This DX molecule gave the structural rigidity required to create a dynamic molecular device, the B-Z switch. DNA double helices can be of three types—the A, B, or the Z-DNA.

The B-DNA is the natural, right-handed helical form of DNA, while the A-DNA is a shrunken, low-humidity form of the B-DNA. Z-DNA can be obtained from certain C-G base repeat sequences occurring in B-DNA that can take a left-handed double helical form. The CG repeated base pair regions can be switched between the left and the right-handed conformations by changing ionic concentration.

The switch was designed in such a way that it had three cyclic strands of DNA, two of them wrapped around a central strand that had the CG-repeat region in the middle. On the two free ends of the side strands fluorescent dyes were attached in order to monitor the conformational change. With the change in ionic concentration the central CG-repeat sequence could alternate between the B and the Z modes bidirectionally which was observed through Forster resonance energy transfer (FRET) spectroscopy.

DNA tweezers

In 2000, Dr. Bernard Yurke and colleagues made an artificial DNA based molecular machine that also accepted DNA as a fuel. The machine, called DNA tweezers, consisted of three strands of DNA labelled A, B, and C. Strands B and C are partially hybridized on to the central strand A with overhangs on both ends. This conformation of the machine is the open conformation. When an auxiliary fuel strand F is introduced, that is designed to hybridize with both overhang regions, the machine attains a closed conformation.

The fuel strand is then removed from the system by the introduction of its exact complement, leaving the system to go back to its original open conformation. This way, a reversible motion is produced, which can be observed by attaching fluorescent tags to the two ends of the strand A. In this case the 5′end was labelled with the dye TET (tetrachloro-.uorescein phosphoramidite) and the 3′ end was labelled with TAMRA (carboxy-tertamethylrhodamine). Aside from the creation of a completely new molecular machine, this showed a way of selective fueling of such machines.

The fuel strands are sequence-specific, hence they will work on only those machines towards which they are directed, and not trigger other machines surrounding them. This machine was later improved to forma Three-State Device which had two robust states and one flexible intermediate state. A variation of the tweezers came about as the DNA-scissors.

Inorganic Molecular Machines

In the past two decades, chemists have been able to create, modify, and control many different types of molecular machines. Many of these machines carry a striking resemblance with the everyday macroscale machines such as gears, propellers, shuttles, etc. Not only this, all of these machines are easy to synthesize artificially and are generally more robust than the natural molecular machines.

Most of these machines are organic compounds of carbon, nitrogen, and hydrogen, with the presence of a metal ion being required occasionally. Electrostatic interactions, covalent and hydrogen bonding play essential role in the performance of these machines. Such artificial chemical machines are controllable in various ways - chemically, electrochemically, and photochemically.

Some of them are even controllable by more than one way, rendering more flexibility and enhancing their utility. A scientist can have more freedom with respect to the design of chemical molecular machines depending on the performance requirements and conditions.

Rotaxanes

The Rotaxane family of molecular machines is characterised by two parts-adumbbell shaped compound with two heavy chemical groups at the ends and a light, cyclic component, called macrocycle, interlocked between the heads as shown in Figure 6 It has been shown that a reversible switch can be made with a rotaxane setup.

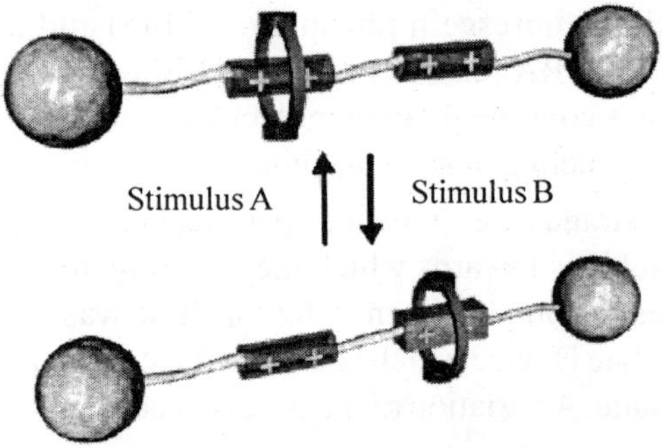

Figure 6. A typical rotaxane shuttle set-up.

For this, one needs to have two chemically active recognition sites in the neck region of the dumbbell. In this particular example, the thread was made of polyether, marked by recognition sites hydroquinol units and terminated at the ends by large triisoproplylsilyl groups.

A tetracationic bead was designed and self-assembled into the system that interacts with the recognition sites. The macrocycle has a natural low energy state on the first recognition site, but can be switched among the two sites reversibly upon application of suitable stimuli.

Depending on the type of rotaxane setup the stimuli can be chemical, electrochemical, or photochemical. The stereo-electronic properties of the recognition sites can be altered by protonation or deprotonation, or by oxidation or reduction, thereby changing their affinity towards the macrocycle. In a recent example, light-induced acceleration of rotaxane motion was achieved by photoisomerisation, while similar controls through alternating current was shown before.

Catenanes

The catenanes are also special type of interlocked structures that represent a growing family of molecular machines. They are synthesized by supramolecular assistance to molecular synthesis. The general structure of a catenane is that of two interlocked ring-like components that are noncovalently linked via a mechanical bond, that is, they are held together without any valence forces. Both the macrocyclic components have recognition sites that are atoms of groups that are redox-active or photochemically reactive. It is possible to have both rings with similar recognition sites.

In such a scenario, one of the rings may rotate inside the other with the conformations stabilised by noncovalent interactions, but the two states of the inner ring differing by 180°. will be undistinguishable. For better control and distinguishable molecular conformations, it is desirable to have different recognition sites within the macrocycles. Then they can be controlled independently through their own specific stimuli.

The stereoelectronic property of one recognition site within a macrocycle can be varied such that at one point it has more affinity to the sites on the other ring. At this instant, the force balance will guide the rotating macrocycle for a stable conformation that requires that particular site to be inside the other macrocycle.

Similarly, with other stimulus, this affinity can be turned off, or even reversed along with the affinity of the second recognition site on the rotating macrocycle increased towards those on the static one. There is a need for computational modelling, simulation, and analysis of such molecular machine motion.

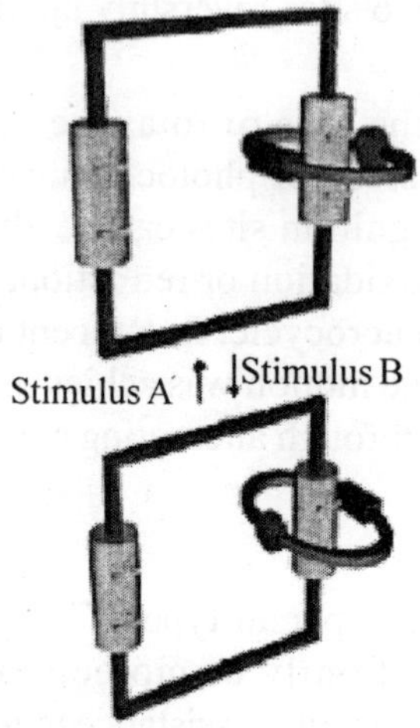

Figure 7. A nondegenerate catenane.

Like rotaxanes, catenanes also can be designed for chemical, photochemical, or electrochemical control. Figure 7 describes one such catenane molecular motor. For both rotaxane and catenane-based molecular machines, it is desirable to have recognition sites such that they can be easily controlled externally. Hence, it is preferable to build sites that are either redox-active or photo-active. Catenanes can also be self-assembled. An example of catenane assembled molecular motors is the electronically controllable bistable switch. An intuitive way of looking at catenanes is to think of them as molecular equivalents of ball and socket and universal joints.

Pseudorotaxanes are structures that contain a ring-like element and a thread-like element that can be "threaded" or "dethreaded" onto the ring upon application of various stimuli. Again, the stimuli can be chemical, photochemical, or electrochemical. These contain a promise of forming molecular machine components analogous to switches and nuts and bolts from the macroscopic world.

Other inorganic molecular machines

Many other molecular devices have been reported in the past four decades that bear a striking resemblance to macroscopic machinery. Chemical compounds behaving as bevel gears and propellers that were reported in the late 1960s and early 1970s are still being studied today.

A molecular propeller can be formed when two bulky rings such as the aryl rings are connected to one central atom, often called the focal atom.

Clockwise rotation of one such ring induces a counterclockwise rotation of the opposite ring about the bond connecting it to the central atom. It is possible to have a three-propeller system as well. Triptycyl and amide ring systems have been shown to observe a coordinated gear-like rotation. "Molecular Turnstiles" which are rotating plates inside amacrocycle, have been created. Such rotations, however, were not controllable. A rotation of a molecular ring about a bond could be controlled by chemical stimuli, and this was shown by Kelly et al. when they demonstrated a molecular brake.

A propeller-like rotation of a 9-triptycyl ring system, which was used in gears, this time connected to a 2, 2′-bipyridine unit could be controlled by the addition and subsequent removal of a metal. Thus, free rotations along single bonds can be stopped and released at will. Another type of molecular switch is the 1 "chiroptical molecular switch". Another large cyclic compound was found to be switchable between its two stable isomeric forms P and M′ (right and left handed) stimulated by light..

Depending on the frequency of light bombarded on it the cis and the trans conformations of the compound 4-[9′ (2′ -methoxythioxanthylidene)] -7-methyl-1,2,3,4-tetrahydrophenanthrene can be interconverted. Allowing a slight variation to this switch, a striking molecular motor driven by light or heat or both was introduced by Koumura et al.. As opposed to the rotation around a single bond in the ratchet described above, the rotation was achieved around a carbon-carbon double bond in a helical alkene.

Ultraviolet light or the change in temperature could trigger a rotation involving four isomerization steps in the compound (3R,3′R)-(P,P)-trans-1,1′, 2, 2′, 3, 3′, 4, 4′-octahydro-3, 3′-dimethyl-4, 4′,-biphenanthrylidene. A second generation motor along with eight other motors from the same material is now operational. This redesigned motor has distinct upper and lower portions and it operates at a higher speed. This motor also provides a good example of how controlled motion at the molecular level can be used to produce a macroscopic change in a system that is visible to the naked eye. The light-driven motors when inside liquid crystal (LC).lms can produce a colour change by inducing a reorganisation of mesogenic molecules.

Nanorobotics Design and Control

Design of Nanorobotic Systems

The design of nanorobotic systems requires the use of information from a vast variety of sciences ranging from quantum molecular dynamics to kinematic analysis. There are many complexities, which are associated with using biocomponents, but the advantages of using these are also quite considerable. These biocomponents offer immense variety and functionality at a scale where creating a man-made material with such capabilities would be extremely difficult. These biocomponents have been perfected by nature through millions of years of evolution and hence these are very accurate and efficient.

F_1-ATPase is known to work at efficiencies which are close to 100%. Such efficiencies, variety, and form are not existent in any other form of material found today. Also, the other significant advantages in using protein-based bio-nanocomponents is the development and refinement over the last 30 years of tools and techniques that enable researchers to mutate proteins in almost any way imaginable. These mutations can consist of anything from simple amino acid sidechain swapping, to amino acid insertions or deletions, incorporation of nonnatural amino acids, and even the combination of unrelated peptide domains into whole new structures.

An excellent example of this approach is the engineering of the F_1-ATPase, which is able to rotate a nanopropeller in the presence of ATP. A computational algorithm was used to determine the mutations necessary to engineer an allosteric zinc-binding site into the F_1-ATPase using site-directed mutagenesis. The mutant F_1-ATPase was then shown to rotate an actin filament in the presence of ATP with average torque of 34 pN nm. This rotation could be stopped with the addition of zinc, and restored with the addition of a chelator to remove the zinc from the allosteric binding site. This type of approach can be used for the improvement of other protein-based nanocomponents.

Hence, these biocomponents seem to be a very logical choice for designing nanorobots. Some of the core applications of nanorobots are in the medical field and using biocomponents for these applications seems to be a good choice as they offer both efficiency and variety of functionality. This idea is clearly inspired by nature's construction of nanorobots, bacteria, and viruses which could be termed as intelligent organisms capable of movement, sensing, and organised control. Hence, the scope would be limited to the usage of these biocomponents in the construction of bio-nanorobotics. A roadmap is proposed which details the main steps towards the design and development of bio-nanorobots.

Roadmap

The roadmap for the development of bio-nanorobotic systems for future applications (medical, space, and military) is shown in

Step 1: Bio-Nanocomponents: Development of bio-nanocomponents from biological systems is the first step towards the design and development of an advanced bio-nanorobot, which could be used for future applications (Figure 8).

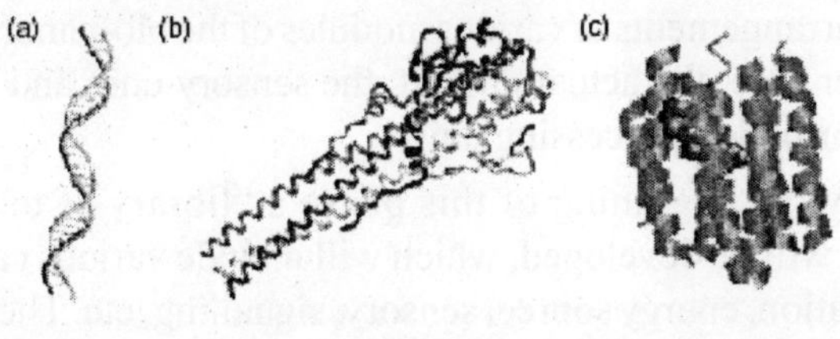

Figure 8. (Step 1) 5 years from now — understanding of basic biological components and controlling their functions as robotic components.

Since the planned systems and devices will be composed of these components, we must have a sound understanding of how these behave and how could they be controlled. From the simple elements such as structural links to more advanced concepts such as motors, each component must be carefully studied and possibly manipulated to understand the functional limits of each one of them.

DNA and carbon nanotubes are being fabricated into various shapes, enabling possibilities of constructing newer and complex devices. These nanostructures are potential candidates for integrating and housing the bionanocomponents within them. Proteins such as rhodopsin and bacteriorhodopsin are a few examples of such bio-nanocomponents.

Both these proteins are naturally found in biological systems as light sensors. They can essentially be used as solar collectors to gather abundant energy from the sun. This energy could either be harvested for later use or could be consumed immediately by other components, such as the ATP Synthase nanorotary motor.

The initial work is intended to be on the biosensors, such as, heat shock factor. These sensors will form an integral part of the proposed bio-nanoassemblies, where these will be integrated within a nanostructure and will get activated, as programmed, for gathering the required information at the nano scale. Tools and techniques from molecular modelling and protein engineering will be used to design these modular components.

Step 2: Assembled Bio-Nanorobots: The next step involves the assembly of functionally stable bio-nanocomponents into complex assemblies. The modular organisation defines the hierarchy rules and spatial arrangements of various modules of the bio-nanorobots such as: the inner core; the actuation unit; the sensory unit; and the signalling and information processing unit.

By the beginning of this phase a "library of bio-nanocomponents" will be developed, which will include various categories such as, actuation, energy source, sensory, signalling, etc. Thereon, one will be able to design and develop such bio-nanosystems that will have enhanced mobile characteristics, and will be able to transport themselves as well as other objects to desired locations at nano scale.

Furthermore, some bio-nanorobots will have the capability of assembling various biocomponents and nanostructures from in situ resources to house fabrication sites and storage areas, while others will just manipulate existing structures by repairing damaged walls or making other renovations.

There will also be robots that not only perform physical labor, but also sense the environment and react accordingly. There will be systems that will sense an oxygen deprivation and stimulate other components to generate oxygen, creating an environment with stable homoeostasis.

Step 3: Distributive Intelligence Programming and Control: With the individual bio-nanorobots in full function, they will now need to collaborate with one another to further develop systems and "colonies" of similar and diverse nanorobots. This step will lay the foundation to the concept of bio-nanoswarms (Figure 9[a]).

Figure 9. (Step 3) (a) Basic bio-nanorobot forming a small swarm of five robots. (b) A basic bio-nanocomputational cell.

Here work has to be performed towards control and programming of bio-nanoswarms. This will evolve concepts like distributive intelligence in the context of bio-nanorobots. Designing swarms of bio-nanorobots capable of carrying out complex tasks and capable of computing and collaborating amongst the group will be the focus. Therefore, the basic computational architectures need to be developed and rules need to be evolved for the bio-nanorobots to make intelligent decisions at the nano scale.

To establish an interface with the macroworld, the computers and electronic hardware have to be designed as well. Figure 10 shows the overall electronic communication architecture. From a location, humans should be able to control and monitor the behavior and action of these swarms. Also, the basic computational capabilities required for functioning of the swarms will be developed. A representative computational bio-nanocell, which will be deployed within a bio-nanorobot, is shown in Figure 9(b).

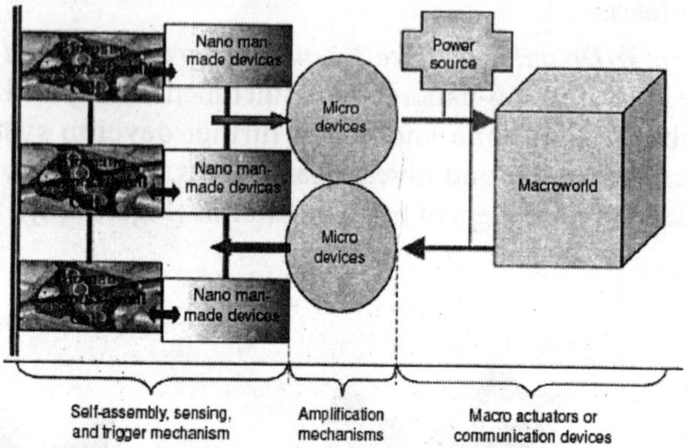

Figure 10. Feedback path from nano to macroworld route.

This basic computational cell will initially be designed for data retrieval and storage at the nano scale. This capability will enable us to programme (within certain degrees of freedom) the swarm behavior in the bio-nanorobots. We will further be able to get their sensory data (from nanoworld) back to the macroworld through these storage devices. This programming capability would form the core essence of a bio-nanorobotics system and hence enables them with immense power.

Step 4: Automatic Fabrication and Information Processing Machines: For carrying out complex missions, such as sensing, signalling, and storing, colonies of these bionanorobotic swarms need to be created. The next step in nanorobotic designing would see the emergence of automatic fabrication methodologies of such bio-nanorobots in vivo and in vitro.

Capability of information processing, which will involve learning and decision making abilities, will be a key consideration of this step. This would enable bioswarms to have capability of self-evolving based on the environment they will be subjected to. These swarms could be programmed to search for alternate energy sources and would have an ability to adapt as per that resource. Energy management, self-repairing, and evolving will be some of the characteristics of these swarms.

Design philosophy for the bio-nanorobotic systems

Modular organisation

Modular organisation defines the fundamental rule and hierarchy for constructing a bio-nanorobotic system. Such construction is performed through stable integration of the individual "bio-modules or components," which constitute the bio-nanorobot. For example, if the entity ABCD, defines a bio-nanorobot having some functional specificity then, A, B, C, and D are said to be the basic biomodules defining it.

The basic construction will be based on the techniques of molecular modelling with emphasis on principles such as Energy Minimization on the hyper surfaces of the biomodules; Hybrid Quantum-Mechanical and Molecular Mechanical methods; Empirical Force field methods; and Maximum Entropy Production in least time.

Modular organisation also enables the bio-nanorobots with capabilities such as, organising into swarms, a feature, which is extremely desirable for various applications. A particular module could consist of other group of modules, just like a fractal structure. The concept of bio-nanocode has been devised, which basically describes the unique functionality of a bio-nanocomponent in terms of alphabetic codes. Each bio-nanocode represents a particular module defining the structure of the bio-nanorobot.

For instance, a code like E-M-S will describe a bio-nanorobot having capabilities of energy storage, mechanical actuation, and signalling at the nano scale. Such representations will help in general classifications and representative mathematics of bio-nanorobots and their swarms.

Universal template-bio-nanostem system

The modular construction concept involves designing a universal template for bio-nanosystems, which could be "programmed and grown" into any possible bio-nanocoded system. This concept mimics the embryonic stem cells found in the human beings, that are a kind of primitive human cells which give rise to all other specialised tissues found in a human fetus, and ultimately to all the three trillion cells in an adult human body.

Our bio-nanostem system will act in a similar way. This universal growth template will be constituted of some basic bio-nanocodes, which will define the bio-nanostem system. This stem system will be designed in a manner that could enable it to be programmed at run-time to any other required biomodule.

Studying bio-nanorobotic systems

Computational methods

Molecular modelling techniques in sync with extensive experimentations will form the basis for studying bio-nanorobotic systems. Some of the molecular modelling techniques that can be used are described below.

Empirical Force Field Methods: (Molecular Mechanics)—In this method, the motion of the electrons is ignored and the energy of the system is calculated based on the position of the nucleus in a particular molecular configuration. There are several approximations which are used in this method with the very basic being the Born-Oppenheimer. The modelling of bio-nanocomponents or their assemblies could be done using one of the energy functions:

$$V(r^N) = \sum_{bonds} \frac{k_i}{2}(l_i - l_{i,0})^2 + \sum_{angles} \frac{k_i}{2}(\theta_i - \theta_{i,0})^2 + \sum_{torsions} \frac{V_n}{2}(1 + \cos(nw - \gamma))$$

$$+ \sum_{i=1}^{N} \sum_{j=i+1}^{N} \left(4\varepsilon ij \left[\left(\frac{\sigma ij}{rij} \right)^{12} - \left(\frac{\sigma ij}{rij} \right)^{6} \right] + \frac{qiqj}{4\pi\varepsilon_0 rij} \right)$$

where $V(r^N)$ denotes the potential energy of N particles having position vector (r).

Energy Minimization Methods: The potential energy depends upon the coordinates of the molecular configuration considered. In this method, points on the hypersurface (potential energy surface) are calculated for which the function has the minimum value. Such geometries, which correspond to the minimum energy, are the stable states of the molecular configuration considered.

By this method we can also analyse the change in the configuration of the system from one minimum state to another. Methods such as the Newton-Raphson and Quasi-Newton are employed to calculate these minima. This method of finding minimum energy points in the molecule is used to prepare for other advanced calculations such as molecular dynamics or Monte-Carlo simulations. It is further used to predict various properties of the system under study.

Molecular Dynamics Simulation Methods: To begin predicting the dynamic performance of a biocomponent (say a peptide) molecular dynamics (MD) is performed. This method utilises Newton's Law of motion through the successive configuration of the system to determine its dynamics. Another variant of molecular dynamics employed in the industry is the Monte-Carlo simulation, which utilises stochastic approaches to generate the required configuration of a system. Simulations are performed based on the calculation of the free energy that is released during the transition from one configuration state to the other.

In MD, the feasibility of a particular conformation of a biomolecule is dictated by the energy constraints. Hence, a transition from one given state to another must be energetically favourable, unless there is an external impetus that helps the molecule overcome the energy barrier. When a macromolecule changes conformation, the interactions of its individual atoms with each other—as well as with the solvent—constitute a very complex force system. With one of the aspects of CHARMM, it is possible to model say, a peptide based on its amino acid sequence and allow a transition between two known states of the protein using targeted molecular dynamics (TMD). TMD is used for approximate modelling of processes spanning long time-scales and relatively large displacements.

Molecular Kinematic Simulations: They are also being used to study the geometric properties and conformational space of the biomolecules (peptides). The kinematic analysis is based on the development of direct and inverse kinematic models and their use towards the workspace analysis of the biomolecules. This computational study calculates all geometrically feasible conformations of the biomolecule, that is, all conformations that can be achieved without any atom interference. This analysis suggests the geometric paths that could be followed by a biomolecule during the transition from the initial to a.nal state, while molecular dynamics narrow downs the possibilities by identifying the only energetically feasible paths. The workspace analysis also characterises the geometrically feasible workspace in terms of dexterity.

Modular Pattern Recognition and Clustering Function: The instantaneous value of the property A of a molecular system can be written as $A(p^n(t), r^n(t))$, where $p^n(t)$ and $r^n(t)$ represent the n momenta and positions of a molecular system at time t. This instantaneous value would vary with respect to the interactions between the particles. There are two parts to this function (termed as M function). One part (function A), evaluates and hence recognises the equivalent modules (in term of properties) in the molecular system.

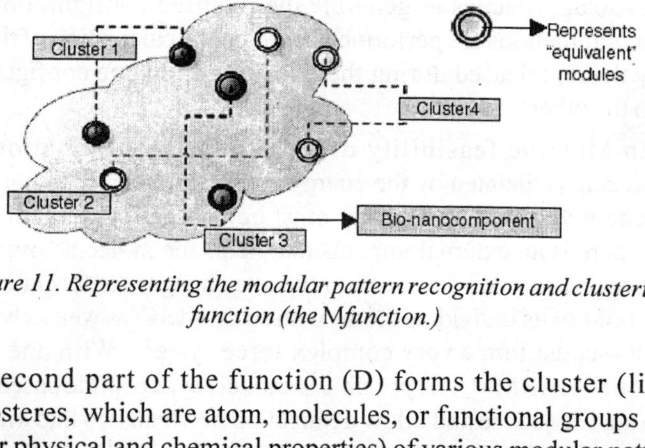

Figure 11. Representing the modular pattern recognition and clustering function (the M*function.)*

The second part of the function (D) forms the cluster (like a bioisosteres, which are atom, molecules, or functional groups with similar physical and chemical properties) of various modular patterns according to the characteristic behaviours as identified by function A and also tracks their variations with rest to time.

$$M \equiv \{A(p^n(t), r^n(t), C^n[n]), D(A(p, r, C, t), f(x, y,...), Bn(t))\}$$

where, A is the first part of the function; $p^n(t)$ and $r^n(t)$ represents the n momenta and positions of a molecular system at time t ; and $C^n[n]$ is an n-dimensional matrix element for the individual components of the molecular system, which would store the categorised values of the modular patterns. D is the second part of the function, and it takes on function A recursively. It also maps these clusters based on a fitness function and stores the time-variant value in subfunction B. Figure 11, represents the conceptual version of the M-function.

This function is based on the hypothesis that in a complex molecular system, there are certain parts, which have similar properties and behavior as the system goes from one state to the other. By identifying these property patterns we can considerably reduce the simulation and computational time frames and can have better predictability of the system. Having established the design for the bionanocomponents, computational studies would focus on determining the overall design and architecture of the bio-nanoassemblies. One of the methods used in the industry is of molecular docking.

Molecular Docking and Scoring Functions: The molecular docking method is very important for the design of nanorobotic systems. This method is utilised to fit two molecules together in 3D space. Themethod of molecular docking could be used to computationally design the bio-nanoassemblies. The algorithms like DOCK, genetic algorithms and distance geometry are being specifically explored.

Many drug-design companies are using depth-first systematic conformational search algorithms for docking, and therefore, is a primary candidate for current research. The docking algorithm generates a large number of solutions. Therefore, we require scoring functions to re fine the results so produced. Scoring functions in use today approximate the "binding free energy" of the molecules. The free energy of binding can be written as an additive equation of various components reflecting the various contributions to binding. A typical equation would be:

$$\Delta G_{bind} = \Delta G_{solvent} + \Delta G_{cont} + G_{\text{int}} + \Delta G_{rot} + \Delta G_{t/r} + \Delta G_{Vib}$$

where, $G_{solvent}$ is the solvent effects contribution, G_{conf} is due to conformational changes in the biomolecule, G_{int} due to specific molecular interactions, G_{rot} due to restrictive internal rotation of the binding bio-molecules, $G_{t/r}$ is due to loss in the translational and rotational free energies due to binding, and G_{vib} is due to vibrational mode variations. Once the overall architecture to bind the bionanocomponents is carried out through molecular docking, we will need to combine the components into real assemblies.

Connecting Bio-nanocomponents in a Binding Site: There are typically two approaches to design connectors for binding the bio-nanocomponents. One method searches the molecular database and selects a connector based on geometrical parameters. CAVEAT is a traditional routine, which employs this method. The other approach is to design a connector from scratch atom by atom. The geometrical parameters being known these connectors are designed.

A typical method employs designing molecular templates for defining a basic connector for binding the bio-nanocomponents. Feasibility of designing these connectors in real life is one challenge faced by molecular modellers. The minimum energy criterion is employed via the Metropolis Monte-Carlo simulation annealing method to select the basic structure of the final connector. Genetic algorithm approaches are also employed to generate such designs.

References

Farrell C.M., Mackey A.T., Klumpp L.M., and Gilbert S.P., "The role of ATP hydrolysis for kinesin processivity", *J. Biol. Chem.* 277: 17079–17087, 2002.

Howard J., Hudspeth A.J., and Vale R.D., "Movement of microtubules by single kinesin molecules", *Nature* 342: 154–158, 1989.

Kinosita K., Jr., Yasuda R., Noji H., Adachi K., "A rotary molecular motor that can work at near 100% ef.ciency", *Philos. Trans.: Biol. Sci.* 355 (1396): 473–489, 2000.

LymnR.W. and Taylor E.W., "Mechanism of adenosine triphosphate hydrolysis by actomyosin", *Biochemistry* 10: 4617–4624, 1971.

Vale R., "Switches, latches, and ampli.ers: common themes of G proteins and molecular motors", *J. Cell Biol.* 135: 291–302, 1996.

4

Medicinal Nanomaterials

Annual U.S. federal funding for nanotechnology R&D exceeded $500 million in 2002 reaching $849 million in FY 2004 and could approach $1 billion in next year's budget. The European Commission has set aside 1.3 billion euros for nanotechnology research during 2003-2006, with annual nanotechnology investment worldwide reaching approximately $3 billion in 2003. The worldwide market for nanoscale devices and molecular modelling should grow 28%/year, rising from $406 million in 2002 to $1.37 billion in 2007, with a 35%/year growth rate in revenues from biomedical nanoscale devices.

In December 2002 the U.S. National Institutes of Health announced a 4-year programme for nanoscience and nanotechnology in medicine. Burgeoning interest in the medical applications of nanotechnology has led to the emergence of a new field called nanomedicine. Most broadly, nanomedicine is the process of diagnosing, treating, and preventing disease and traumatic injury, of relieving pain, and of preserving and improving human health, using molecular tools and molecular knowledge of the human body.

The NIH Roadmap's new Nanomedicine Initiatives, first released in late 2003, "envision that this cutting-edge area of research will begin yielding medical benefits as early as 10 years from now" and will begin with "establishing a handful of Nanomedicine Centers...staffed by a highly interdisciplinary scientific crew including biologists, physicians, mathematicians, engineers and computer scientists...gathering extensive information about how molecular machines are built" who will also develop "a new kind of vocabulary-lexicon- to define biological parts and processes in engineering terms".

Even state-funded programmes have begun, such as New York's Alliance for Nanomedical Technologies. In the relatively near term, over the next 5 years, nanomedicine can address many important medical problems by using nanoscale-structured materials and simple nanodevices that can be manufactured today. This includes the interaction of nanostructured materials with biological systems.

Over the next 5-10 years, biotechnology will make possible even more remarkable advances in molecular medicine and biobotics-microbiological robots or engineered organisms. In the longer term, perhaps 10-20 years from today, the earliest molecular machine systems and nanorobots may join the medical armamentarium, finally giving physicians the most potent tools imaginable to conquer human disease, ill-health, and aging.

Nanodevices and Medical Nanomaterials

Role of Nanopores

Perhaps one of the simplest medical nanomaterials is a surface perforated with holes, or nanopores. In 1997 Desai and Ferrari created what could be considered one of the earliest therapeutically useful nanomedical devices, employing bulk micromachining to fabricate tiny cell-containing chambers within single crystalline silicon wafers. The chambers interface with the surrounding biological environment through polycrystalline silicon.

Filter membranes which are micromachined to present a high density of uniform nanopores as small as 20 nanometers in diameter. These pores are large enough to allow small molecules such as oxygen, glucose, and insulin to pass, but are small enough to impede the passage of much larger immune system molecules such as immunoglobulins and graft-borne virus particles.

Safely ensconced behind this artificial barrier, immunoisolated encapsulated rat pancreatic cells may receive nutrients and remain healthy for weeks, secreting insulin back out through the pores while the immune system remains unaware of the foreign cells which it would normally attack and reject.

Microcapsules containing replacement islets of Langerhans cells- most likely easily-harvested piglet islet cells-could be implanted beneath the skin of some diabetes patients. This could temporarily restore the body's delicate glucose control feedback loop without the need for powerful immunosuppressants that can leave the patient at serious risk for infection. Supplying encapsulated new cells to the body could also be a valuable way to treat other enzyme or hormone deficiency diseases, including encapsulated neurons which could be implanted in the brain and then be electrically stimulated to release neurotransmitters, possibly as part of a future treatment for Alzheimer's or Parkinson's diseases. The flow of materials through nanopores can also be externally regulated.

The first artificial voltage-gated molecular nanosieve was fabricated by Martin and colleaguesin 1995. Martin's membrane contains an array of cylindrical gold nanotubules with inside diameters as small as 1.6 nanometers. When the tubules are positively charged, positive ions are excluded and only negative ions are transported through the membrane. When the membrane receives a negative voltage, only positive ions can pass. Future similar nanodevices may combine voltage gating with pore size, shape, and charge constraints to achieve precise control of ion transport with significant molecular specificity.

Martin's recent efforts have been directed at immobilising biochemical molecular-recognition agents such as enzymes, antibodies, other proteins and DNA inside the nanotubes as active biological nanosensors, to perform drug separations, and to allow selected biocatalysis. Others are investigating synthetic nanoporeion pumps, voltage-gated nanopores embedded in artificial membranes, and an ion channel switch biosensor that detects changes in chemical concentration of $\sim 10^{-18}$. Molecular dynamics theoretical studies of viscosity and diffusion through nanopores are in progress.

Finally, Daniel Branton's team at Harvard University has conducted an ongoing series of experiments using an electric field to drive a variety of RNA and DNA polymers through the central nanopore of an alpha-hemolysin protein channel mounted in a lipid bilayer similar to the outer membrane of a living cell. By 1998, Branton had shown that the nanopore could be used to rapidly discriminate between pyrimidine and purine segments along a single RNA molecule. In 2000, the scientists demonstrated the ability to distinguish between DNA chains of similar length and composition that differ only in base pair sequence.

Current research is directed toward reliably fabricating pores with specific diameters and repeatable geometries at high precision, understanding the unzipping of double-stranded DNA as one strand is pulled through the pore and the recognition of folded DNA molecules passing through the pore, experiments with new 3-10 nm silicon-nitride nanopores, and investigating the benefits of adding electrically conducting electrodes to pores to improve longitudinal resolution "possibly to the single-base level for DNA". Nanopore-based DNA sequencing devices could allow per-pore read rates potentially up to 1000 bases per second, possibly eventually providing a low-cost high-throughput method for very rapid genome sequencing.

Molecular Imprinting

Another early goal of nanomedicine is to study how biological molecular receptors work, and then to build artificial binding sites on a made-to-order basis to achieve specific medical results. Molecular imprinting is an existing technique in which a cocktail of functionalised monomers interacts reversibly with a target molecule using only noncovalent forces. The complex is then cross-linked and polymerised in a casting procedure, leaving behind a polymer with recognition sites complementary to the target molecule in both shape and functionality. Each such site constitutes an induced molecular "memory," capable of selectively binding the target species.

In one experiment involving an amino acid derivative target, one artificial binding site per (3.8 nm) polymer block was created. Chiral separations, enzymatic transition state activity, and high receptor affinities have been demonstrated. Molecularly imprinted polymers could be medically useful in clinical applications such as controlled drug release, drug monitoring devices, quick biochemical separations and assays, recognition elements in biosensors and chemosensors, and biological and receptor mimics including artificial antibodies (plastibodies) or biomimicking enzymes (plastizymes).

But molecularly imprinted polymers have limitations, such as incomplete template removal, broad guest affinities and selectivities, and slow mass transfer. Imprinting inside dendrimers may allow quantitative template removal, nearly homogeneous binding sites, solubility in common organic solvents, and amenability to the incorporation of o'her functional groups.

Quantum Dots Nanocrystals

Fluorescent tags are commonplace in medicine and biology, found in everything from HIV tests to experiments that image the inner functions of cells. But different dye molecules must be used for each colour, colour-matched lasers are needed to get each dye to fluoresce, and dye colours tend to bleed together and fade quickly after one use. "Quantum dot" nanocrystals have none of these shortcomings. These dots are tiny particles measuring only a few nanometers across, about the same size as a protein molecule or a short sequence of DNA.

They come in a nearly unlimited palette of sharply defined colours which can be customised by changing particle size or composition. Particles can be excited to fluorescence with white light, can be linked to biomolecules to form long-lived sensitive probes to identify specific compounds up to a thousand times brighter than conventional dyes used in many biological tests, and can track biological events by simultaneously tagging each biological component with nanodots of a specific colour. Quantum Dot Corp., the manufacturer, believes this kind of flexibility could offer a cheap and easy way to screen a blood sample for the presence of a number of different viruses at the same time.

It could also give physicians a fast diagnostic tool to detect, say, the presence of a particular set of proteins that strongly indicates a person is having a heart attack or to detect known cellular cancer markers. On the research front, the ability to simultaneously tag multiple biomolecules both on and inside cells could allow scientists to watch the complex cellular changes and events associated with disease, providing valuable clues for the development of future pharmaceuticals and therapeutics. Quantum dots are useful for studying genes, proteins and drug targets in single cells, tissue specimens, and living animals.

Quantum dots are being investigated as chemical sensors, for cancer cell detection, gene expression studies, gene mapping and DNA microarray analysis, immunocytochemical probes, intracellular organelle markers, live cell labelling, medical diagnostics and drug screening, SNP (Single Nucleotide Polymorphism) genotyping, vascular imaging, and many other applications. Quantum dot physics has been studied theoretically and computationally using time-dependent density functional theory and other methods.

Researchers from Northwestern University and Argonne National Laboratory have created a hybrid "nanodevice" composed of 4.5-nm nanocrystals of biocompatible titanium dioxide semiconductor covalently attached with snippets of oligonucleotide DNA. Experiments showed that these nanocomposites not only retain the intrinsic photocatalytic capacity of TiO2 and the bioactivity of the oligonucleotide DNA, but more importantly also possess the unique property of a light-inducible nucleic acid endonuclease (separating when exposed to light or x-rays). For example, researchers would attach to the semiconductor scaffolding a strand of DNA that matches a defective gene within a cell, then introduce the nanoparticle into the cell nucleus where the attached DNA binds with its defective complementary DNA strand, whereupon exposure of the bound nanoparticle to light or x-rays snips off the defective gene.

Other molecules besides oligonucleotides can be attached to the titanium dioxide scaffolding, such as navigational peptides or proteins, which, like viral vectors, can help the nanoparticles home in on the cell nucleus. This simple nanocrystal nanodevice might one day be used to target defective genes that play a role in cancer, neurological disease and other conditions, though testing in a laboratory model is at least two years away.

Fullerenes and Nanotubes

Soluble derivatives of fullerenes such as C_{60} have shown great utility as pharmaceutical agents. These derivatives, many already in clinical trials, have good biocompatibility and low toxicity even at relatively high dosages. Fullerene compounds may serve as antiviral agents, antibacterial agents, photodynamic antitumour and anticancertherapies, antioxidants and anti-apoptosis agents which may include treatments for amyotrophic lateral sclerosis and Parkinson's disease. Single-walled and multi-walledcarbon nanotubes are being investigated as biosensors, for example to detect glucose, ethanol, hydrogen peroxide, selected proteins such as immunoglobulins, and as an electrochemical DNA hybridization biosensor.

Nanoshells and Magnetic Nanoprobes

Halas and West at Rice University in Houston have developed a platform for nanoscale drug delivery called the nanoshell. Unlike carbon fullerenes, the slightly larger nanoshells are dielectric-metal nanospheres with a core of silica and a gold coating, whose optical resonance is a function of the relative size of the constituent layers. The nanoshells are embedded in a drug-containing tumour targeted hydrogel polymer and injected into the body. The shells circulate through the body until they accumulate near tumour cells.

When heated with an infrared laser, the nanoshells selectively absorb the IR frequencies, melt the polymer and release their drug payload at a specific site. Nanoshells offer advantages over traditional cancer treatments: earlier detection, more detailed imaging, fast noninvasive imaging, and integrated detection and treatment. This technique could also prove useful in treating diabetes.

Instead of taking an injection of insulin, a patient would use a ballpoint-pen-size infrared laser to heat the skin where the nanoshell polymer had been injected. The heat from nanoshells would cause the polymer to release a pulse of insulin. Unlike injections, which are taken several times a day, the nanoshell-polymer system could remain in the body for months.

Nanospectra Biosciences, a private company started by Halas and West, is developing commercial applications of nanoshell technology. Nanospectra is conducting animal studies at the MD Anderson Cancer at the University of Texas, specifically targeting micrometastases, tiny aggregates of cancer cells too small for surgeons to find and remove ith a scalpel. The company hopes to start clinical trials for the cancer treatment by 2004 and for the insulin delivery system by 2006. In mid-2003, Rice researchers announced the development of a point-of-care whole blood immunoassay using antibody-nanoparticle conjugates of gold nanoshells.

Varying the thickness of the metal shell allow precise tuning of the colour of light to which the nanoshells respond; near-infrared light penetrates whole blood very well, so it is an optimal wavelength for whole blood immunoassay. Successful detection of sub-nanogram-per-millilitre quantities of immunoglobulins was achieved in saline, serum, and whole blood in 10-30 minutes.

An alternative approach pursued by Triton BioSystems is to bond iron nanoparticles and monoclonal antibodies into nanobioprobes about 40 nanometers long. The chemically inert probes are injected and circulate inside the body, whereupon the antibodies selectively bind to tumour cell membranes.

Once the tumour is covered with bioprobes after several hours, a magnetic field generated from a portable alternating magnetic field machine heats the iron particles to more than 170 degrees, killing the tumour cells in a few seconds.

Once the cells are destroyed, the body's excretion system removes cellular residue and nanoparticles alike. Test subjects feel no pain from the heat generated. Triton BioSystems plans to start designing human tests and ask the FDA for permission to begin human clinical trials in 2006.

Mirkin's group at Northwestern University uses magnetic microparticle probes coated with target proteinbinding antibodies plus 13-nm nanoparticle probes with a similar coating but including a unique hybridized "barcode" DNA sequence as an ultrasensitive method for detecting protein analyses such as prostate-specific antigen (PSA).

After the target protein in the test sample is captured by the microparticles, magnetic separation of the complexed microparticle probes and PSA is followed by dehybridization of the bar-code oligonucleotides on the nanoparticle probe surface, allowing the determination of the presence of PSA by identifying the bar-code sequence released from the nanoparticle probe. Using polymerase chain reaction on the oligonucleotide bar codes allows PSA to be detected at 3 attomolar concentration, about a million times more sensitive than comparable clinically accepted conventional assays for detecting the same protein target.

Targeted Nanoparticles

Multisegment gold/nickel nanorods are being explored by Leong's group at Johns Hopkins School of Medicineas tissue-targeted carriers for gene delivery into cells that "can simultaneously bind compacted DNA plasmids and targeting ligands in a spatially defined manner" and allow "precise control of composition, size and multifunctionality of the gene-delivery system." The nanorods are electrodeposited into the cylindrical 100 nm diameter pores of an alumina membrane, joining a 100 nm length gold segment and a 100 nm length nickel segment.

After the alumina template is etched away, the nanorods are functionalised by attaching DNA plasmids to the nickel segments and transferring, a cell-targeting protein, to the gold segments, using molecular linkages that selectively bind to only one metal and thus impart biofunctionality to the nanorods in a spatially defined manner. Leong notes that extra segments could be added to the nanorods, for example to bind additional biofunctionalities such as an endosomolytic agent, or magnetic segments could be added to allow manipulating the nanorods with an external magnetic field.

Targeted radioimmunotherapeutic agentsinclude the FDA-approved "cancer smart bombs" that deliver tumour killing radioactive yttrium or iodine attached to a lymphoma-targeted antibody. Other antibody-linked agents are being investigated such as the alpha-emitting actinium-based "nanogenerator" molecules that use internalising monoclonal antibodies to penetrate the cell and have been shown, in vitro, to specifically kill leukaemia, lymphoma, breast, ovarian, neuroblastoma, and prostate cancer cells at becquerel (picocurie) levels, with promising preliminary results against advanced ovarian cancer in mice. However, drug specificity is still no better than the targeting accuracy of the chosen antibody, and there is significant mistargeting, leading to unwanted side effects.

Enzyme-activated drugs, first developed in the 1980s and still under active investigation, separate the targeting and activation functions. For instance, an antibody directed enzyme-triggered prodrug cancer therapy is being developed by researchers at the University of Gottingen in Germany. This targeted drug molecule turns lethal only when it reaches cancer cells while remaining harmless inside healthy cells. In tests, mice previously implanted with human tumours are given an activating targeted enzyme that sticks only to human tumour cells, mostly ignoring healthy mouse cells. Then the antitumour molecule is injected.

In its activated state, this fungal-derived antibiotic molecule is a highly-strained ring of three carbon atoms that is apt to burst open, becoming a reactive molecule that wreaks havoc among the nucleic acid molecules essential for normal cell function. But the molecule is injected as a prodrug-an antibiotic lacking the strained ring and with a sugar safety-catch. Once the sugar is clipped off by the previously positioned targeted enzyme, the drug molecule rearranges itself into a three-atom ring, becoming lethally active. Notes chemist Philip Ball: "The selectivity of the damage still depends on antibody's ability to hook onto the right cells, and on the absence of other enzymes in the body that also activate the prodrug".

A further improvement in enzyme-activated drugs are "smart drugs" that become medically active only in specific circumstances and in an inherently localised manner. Yoshihisa Suzuki at Kyoto University has designed a novel drug molecule that releases antibiotic only in the presence of an infection. Suzuki started with the common antibiotic molecule gentamicin and bound it to a hydrogel using a newly developed peptide linker. The linker can be cleaved by a proteinase enzyme manufactured by Pseudomonas aeruginosa, a Gram-negative bacillus that causes inflammation and urinary tract infection, folliculitis, and otitis externa in humans.

Tests on rats show that when the hydrogel is applied to a wound site, the antibiotic is not released if no P. aeruginosa bacteria are present. But if any bacteria of this type are present, then the proteolytic enzyme that the microbes naturally produce cleaves the linker and the gentamicin is released, killing the bacteria." If the proteinase specific to each bacterium [species] can be used for the signal," wrote Suzuki, "different spectra of antibiotics could be released from the same dressing material, depending on the strain of bacterium".

In subsequent work an alternative antibiotic release system triggered by thrombin activity, which accompanies Staphylococcus aureus wound infections, was successfully tested as a high-specificity stimulus-responsive controlled drug release system. Other stimulus-responsive "smart" hydrogels are being studied, including a hydrogel composite membrane co-loaded with insulin and glucose oxidase enzyme that exhibits a twofold increase in insulin release rate when immersed in glucose solution, demonstrating "chemically stimulated controlled release" and "the potential of such systems to function as a chemically synthesized artificial pancreas".

Nanoparticles with an even greater range of action are being developed by Raoul Kopelman's group at the University of Michigan. Their current goal is the development of novel molecular nanodevices for the early detection and therapy of brain cancer, using silica-coated iron oxide nanoparticles with a biocompatible polyethylene glycol coating.

The miniscule size of the particles-20-200 nanometers-should allow them to penetrate into areas of the brain that would otherwise be severely damaged by invasive surgery. The particles are attached to a cancer cell antibody or other tracer molecule that adheres to cancer cells, and are affixed with a nanopacket of contrast agent that makes the particles highly visible during magnetic resonance imaging (MRI).

The particles also enhance the killing effect during the subsequent laser irradiation of brain tissue, concentrating the destructive effect only on sick cells unlike traditional chemotherapy and radiation which kills cancerous cells but also destroys healthy cells. Nanoparticles allow MRI to see a few small brain tumour cells as small as 50 microns- depending on the cancer type, tumour cells can range from 5-50 microns each and may grow in locations separate from the tumour site, hence are sometimes not visible to surgeons.

Fei Yan, a postdoc in Kopelman's lab, is working on these nanodevices, called the Dynamic Nano-Platform, now being commercialised as therapeutic "nanosomes" under license to Molecular Therapeutics. According to the company, "the nanosome platform provides the core technology with interchangeable components that provide ultimate flexibility in targeting, imaging and treatment of cancer and cardiovascular disease indications.

Dendrimer-Based Devices

Dendrimers represent yet another nanostructured material that may soon find its way into medical therapeutics. Starburst dendrimers are tree-shaped synthetic molecules with a regular branching structure emanating outward from a core that form nanometer by nanometer, with the number of synthetic steps or "generations" dictating the exact size of the particles, typically a few nanometers in spheroidal diameter.

The peripheral layer can be made to form a dense field of molecular groups that serve as hooks for attaching other useful molecules, such as DNA, which can enter cells while avoiding triggering an immune response, unlike viral vectors commonly employed today for transfection.

Upon encountering a living cell, dendrimers of a certain size trigger a process called endocytosis in which the cell's outermost membrane deforms into a tiny bubble, or vesicle. The vesicle encloses the dendrimer which is then admitted into the cell's interior. Once inside, the DNA is released and migrates to the nucleus where it becomes part of the cell's genome.

The technique has been tested on a variety of mammalian cell types and in animal models, though clinical human trials of dendrimer gene therapy remain to be done. Glycodendrimer "nanodecoys" have also been used to trap and deactivate some strains of influenza virus particles.

The glycodendrimers present a surface that mimics the sialic acid groups normally found in the mammalian cell membrane, causing virus particles to adhere to the outer branches of the decoys instead of the natural cells. In July 2003, Starpharma was cleared by the U.S. FDA for human trials of their dendrimer-based anti- HIV microbicide. Their product has been successful in preventing simian-HIV.

Computational simulations have also been done on some dendrimer-based nanoparticles. James Baker's group at the University of Michigan is extending this work to the synthesis of multi-component nanodevices called tecto-dendrimers built up from a number of single-molecule dendrimer components. Tecto-dendrimers have a single core dendrimer surrounded by additional dendrimer modules of different types, each type designed to perform a function necessary to a smart therapeutic nanodevice.

Baker's group has built a library of dendrimeric components from which a combinatorially large number of nanodevices can be synthesized. The initial library contains components which will perform the following tasks:

(1) diseased cell recognition,

(2) diagnosis of disease state,

(3) drug delivery,

(4) reporting location, and

(5) reporting outcome of therapy.

By using this modular architecture, an array of smart therapeutic nanodevices can be created with little effort. For instance, once apoptosis-reporting, contrast-enhancing, and chemotherapeutic-releasing dendrimer modules are made and attached to the core dendrimer, it should be possible to make large quantities of this tecto-dendrimer as a starting material. This framework structure can be customised to fight a particular cancer simply by substituting any one of many possible distinct cancer recognition or "targeting" dendrimers, creating a nanodevice customised to destroy a specific cancer type and no other, while also sparing the healthy normal cells.

In three nanodevices synthesized using an ethylenediamine core polyamidoamine dendrimer of generation 5, with folic acid, fluorescein, and methotrexate covalently attached to the surface to provide targeting, imaging, and intracellular drug delivery capabilities, the "targeted delivery improved the cytotoxic response of the cells to methotrexate 100-fold over free drug." At least a half dozen cancer cell types have already been associated with at least one unique protein which targeting dendrimers could use to identify the cell as cancerous, and as the genomic revolution progresses it is likely that proteins unique to each kind of cancer will be identified, thus allowing Baker to design a recognition dendrimer for each type of cancer.

The same cell-surface protein recognition-targeting strategy could be applied against virus-infected cells and parasites. Molecular modelling has been used to determine optimal dendrimer surface modifications for the function of tecto-dendrimer nanodevices and to suggest surface modifications that improve targeting. NASA and the National Cancer Institute have funded Baker's lab to produce dendrimer-based nanodevices that can detect and report cellular damage due to radiation exposure in astronauts on long-term space missions.

By mid-2002, the lab had built a dendrimeric nanodevice to detect and report the intracellular presence of caspase-3, one of the first enzymes released during cellular suicide or apoptosis, one sign of a radiation-damaged cell. The device includes one component that identifies the dendrimer as a blood sugar so that the nanodevice is readily absorbed into a white blood cell, and a second component using fluorescence resonance energy transfer (FRET) that employs two closely bonded molecules.

Before apoptosis, the FRET system stays bound together and the white cell interior remains dark upon illumination. Once apoptosis begins and caspase-3 is released, the bond is quickly broken and the white blood cell is a wash in fluorescent light. If a retinal scanning device measuring the level of fluorescence inside an astronaut's body reads above a certain baseline, counteracting drugs can be taken.

Radio-Controlled Biomolecules

While there are already many examples of nanocrystals attached to biological systems for biosensing purposes, the same nanoparticles are now being investigated as a means for directly controlling biological processes. Jacobson and colleagues have attached tiny radio-frequency antennas-1.4 nanometer gold nanocrystals of less than 100 atoms-to DNA.

When a ~1 GHz radio-frequency magnetic field is transmitted into the tiny antennas, alternating eddy currents induced in the nanocrystals produce highly localised inductive heating, causing the double stranded DNA to separate into two strands in a matter of seconds in a fully reversible dehybridisation process that leaves neighbouring molecules untouched.

The long-term goal is to apply the antennas to living systems and control DNA via remote electronic switching. This requires attaching gold nanoparticles to specific oligonucleotides which, when added to a sample of DNA, would bind to complementary gene sequences, blocking the activity of those genes and effectively turning them off. Applying the rf magnetic field then heats the gold particles, causing their attached DNA fragments to detach, turning the genes back on. Such a tool could give pharmaceutical researchers a way to simulate the effects of potential drugs which also turn genes on and off. Says

Gerald Joyce: "You can even start to think of differential receivers-different radio receivers that respond differently to different frequencies. By dialing in the right frequency, you can turn on tags on one part of DNA but not other tags."

The gold nanocrystals can be attached to proteins as well as DNA, opening up the possibility of future radio frequency biology electronically controlling more complex biological processes such as enzymatic activity, protein folding and biomolecular assembly. In late 2002, Jacobson announced that his team had achieved electrical control over proteins as well. The researchers separated an RNA-hydrolyzing enzyme called ribonuclease S into two pieces: a large protein segment made up of 104 amino acids and a small 18-aminoacid strand called the S-peptide.

The RNAase enzyme is inactive unless the small strand sits in the mouth of the protein. Jacobson's group linked gold nanoparticles to the end of S-peptide strands and used the particles as a switch to turn the enzyme on and off-in the absence of the rf field, the S-peptides adopt their usual conformation and the RNAase remains active, but with the external rf field switched on, the rapidly spinning nanoparticles prevented the S-peptide from assembling with the larger protein, inactivating the enzyme.

Microscale Biological Viruses and Bacteria

One convenient short-cut to nanorobotics is to engineer natural nanomachine systems-microscale biological viruses and bacteria-to create new, artificial biological devices. Efforts at purely rational virus design are underway but have not yet borne much fruit. For example, Endy et al. computationally simulated the growth rates of bacteriophage T7 mutants with altered genetic element orders and found one new genome permutation that was predicted to allow the phage to grow 31% faster than wild type; unfortunately, experiments failed to confirm the predicted speed-up. Better models are clearly needed. Nevertheless, combinatorial experiments on wild type T7 by others have produced new but immunologically indistinguishable T7 variants which have 12% of their genome deleted and which replicate twice as fast as wild type.

The Synthetic Biology Lab at MIT is building the next generation T7, a bacteriophage with a genome size of about 40 Kbp and 56 genes.

Considerations in the redesign process include: "adding or removing restriction sites to allow for easy manipulation of various parts, reclaiming codon usage, and eliminating parts of the genome that have no apparent function." Young and Douglashave chemically modified the Cowpea chlorotic mottle virus (CCMV) viral protein cage surface to allow engineering of surface-exposed functional groups. This includes the addition of lamanin peptide 11 to the viral coat, and the incorporation of 180 gadolinium atoms into each 28-nm viral capsid, allowing these tumour-targeting particles to serve as tumour-selective MRI contrast agents.

The researchers have investigated re-engineering the artificial virion to make a complete tumour-killing nanodevice, exploiting a gating mechanism that results from reversible structural transitions in the virus. The natural viral gate of CCMV has been re-engineered to allow control by redox potential; cellular interiors have a higher redox potential than blood, so viral capsids could be shut tight in transit but would open their redox-controlled gates after entering targeted cancer cells, releasing their payload of therapeutic agents. In principle, the four capabilities of the engineered capsids- high-sensitivity imaging, cell targeting, drug transport, and controlled delivery-represent a potentially powerful, yet minimally toxic, way to fight metastasized cancer.

The rational design and synthesis of chimeric viral replicators is already possible today, and the rational design and synthesis of completely artificial viral sequences, leading to the manufacture of completely synthetic viral replicators, should eventually be possible. In a three-year projectculminating in 2002, the 7500-base polio virus was rationally manufactured "from scratch" in the laboratory by synthesizing the known viral genetic sequence in DNA, enzymatically creating an RNA copy of the artificial DNA strand, then injecting the synthetic RNA into a cell-free broth containing a mixture of proteins taken from cells, which then directed the synthesis of complete polio virion particles.

Engineered bacterial "biorobots" are also being pursued. Mushe gianconcludes that as few as 300 highly conserved genes are all that may be required for life, constituting the minimum possible genome for a functional microbe. An organism containing this minimal gene set would be able to perform the dozen or so functions required for life-manufacturing cellular biomolecules, generating energy, repairing damage, transporting salts and other molecules, responding to environmental chemical cues, and replicating.

Thus a minimal synthetic microbe- a basic cellular chassis-could be specified by a genome only 150,000 nucleotide bases in length. Used in medicine, these artificial biorobots could be designed to produce useful vitamins, hormones, enzymes or cytokines in which a patient's body was deficient, or to selectively absorb and metabolise into harmless end products harmful substances such as poisons, toxins, or indigestible intracellular detritus, or even to perform useful mechanical tasks.

In November 2002, J.Craig Venter, of human genome sequencing fame, and Hamilton O.Smith, a Nobel laureate, announced that their new company, Institute for Biological Energy Alternatives (IBEA), had received a $3 million, three-year grant from the Energy Department to create a minimalist organism, starting with the Mycoplasma genitalium microorganism. Working with a research staff of 25 people, the scientists are removing all genetic material from the organism, then synthesizing an artificial string of genetic material resembling a naturally occurring chromosome that they hope will contain the minimum number of M. genitalium genes needed to sustain life.

The artificial chromosome will be inserted into the hollowed-out cell, which will then be tested for its ability to survive and reproduce. To ensure safety, the cell will be deliberately hobbled to render it incapable of infecting people, and will be strictly confined and designed to die if it does manage to escape into the environment. In 2003, Egea Biosciences received "the first [patent] to include broad claims for the chemical synthesis of entire genes and networks of genes comprising a genome, the 'operating system' of living organisms." Egea's proprietary GeneWriter™ and Protein Programming™ technology has:

(1) produced libraries of more than 1,000,000 programmed proteins,

(2) produced over 200 synthetic genes and proteins,

(3) produced the largest gene ever chemically synthesized of over 16,000 bases,

(4) engineered proteins for novel functions,

(5) improved protein expression through codon optimization, and

(6) developed custom genes for protein manufacturing in specific host cells.

Egea's software allows researchers to author new DNA sequences that the company's hardware can then manufacture to specification with a base-placement error of only $\sim 10^{-4}$, which Egea calls "word processing for DNA". The patent recites one preferred embodiment of the invention as the synthesis of "a gene of 100,000 bp...from one thousand 100-mers. The overlap between 'pairs' of plus and minus oligonucleotides is 75 bases, leaving a 25 base pair overhang.

In this method, a combinatorial approach is used where corresponding pairs of partially complementary oligonucleotides are hybridized in the first step. A second round of hybridization then is undertaken with appropriately complementary pairs of products from the first round. This process is repeated a total of 10 times, each round of hybridization reducing the number of products by half. Ligation of the products then is performed." The result would be a strand of DNA 100,000 base pairs in length, long enough to make a very simple bacterial genome.

Medical Nanorobotics

The third major development pathway of nanomedicine-molecular nanotechnology (MNT) or nanorobotics- takes as its purview the engineering of complex nanomechanical systems for medical applications. Just as biotechnology extends the range and efficacy of treatment options available from nanomaterials, the advent of molecular nanotechnology will again expand enormously the effectiveness, precision and speed of future medical treatments while at the same time significantly reducing their risk, cost, and invasiveness. MNT will allow doctors to perform direct in vivo surgery on individual human cells. The ability to design, construct, and deploy large numbers of microscopic medical nanorobots will make this possible.

Medical Nanorobotics

In his remarkably prescient 1959 talk "There's Plenty of Room at the

Bottom," the late Nobel physicist Richard P.Fe ynman proposed employing machine tools to make smaller machine tools, these to be used in turn to make still smaller machine tools, and so on all the way down to the atomic level. Feynman was clearly aware of the potential medical applications of the new technology he was proposing.

After discussing his ideas with a colleague, Feynman offered the first known proposal for a nanomedical procedure to cure heart disease: "A friend of mine suggests a very interesting possibility for relatively small machines. He says that, although it is a very wild idea, it would be interesting in surgery if you could swallow the surgeon.

You put the mechanical surgeon inside the blood vessel and it goes into the heart and looks around. It finds out which valve is the faulty one and takes a little knife and slices it out. Other small machines might be permanently incorporated in the body to assist some inadequately functioning organ." Later in his historic lecture in 1959, Feynman urged us to consider the possibility, in connection with biological cells, "that we can manufacture an object that maneuvers at that level!".

The vision behind Feynman's remarks became a serious area of inquiry two decades later, when K. Eric Drexler, while still a graduate student at the Massachusetts Institute of Technology, published a technical paper suggesting that it might be possible to construct, from biological parts, nanodevices that could inspect the cells of a living human being and carry on repairs within them. This was followed a decade later by Drexler's seminal technical booklaying the foundations for molecular machine systems and molecular manufacturing, and subsequently by Freitas' technical bookson medical nanorobotics.

Nanorobot Parts and Components

Nanobearings and nanogears

In order to establish the feasibility of molecular manufacturing, it is first necessary to create and to analyse possible designs for nanoscale mechanical parts that could, in principle, be manufactured. Because these components cannot yet be physically built in 2004, such designs cannot be subjected to rigorous experimental testing and validation. Designers are forced instead to rely upon ab initio structural analysis and molecular dynamics simulations. Notes Drexler: "Our ability to model molecular machines of specific kinds, designed in part for ease of modelling, has far outrun the ability to make them. Design calculations and computational experiments enable the theoretical studies of these devices, independent of the technologies needed to implement them." Molecular bearings are perhaps the most convenient class of components to design because their structure and operation is fairly straightforward.

One of the simplest examples is Drexler's overlap-repulsion bearing design. This bearing has 206 atoms of carbon, silicon, oxygen and hydrogen, and is composed of a small shaft that rotates within a ring sleeve measuring 2.2 nm in diameter.

The atoms of the shaft are arranged in a 6-fold symmetry, while the ring has 14-fold symmetry, a combination that provides low energy barriers to shaft rotation. Figure 1 shows an exploded view of a 2808-atom strained-shell sleeve bearing designed by Drexler and Merkleusing molecular mechanics force fields to ensure that bond lengths, bond angles, van der Waals distances, and strain energies are reasonable. This 4.8-nm diameter bearing features an interlocking-groove interface which derives from a modified diamond (100) surface.

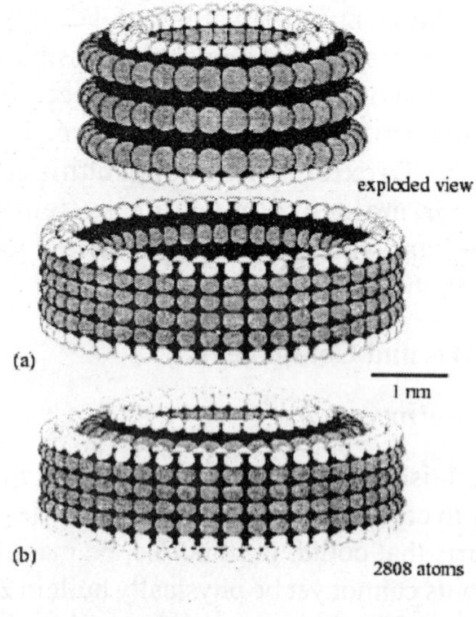

Figure 1. Exploded view of a 2808-atom strained-shell sleeve bearing.

Ridges on the shaft interlock with ridges on the sleeve, making a very stiff structure. Attempts to bob the shaft up or down, or rock it from side to side, or displace it in any direction encounter a very strong resistance. Molecular gears are another convenient component system for molecular manufacturing design-ahead. For example, Drexler and Merkle designed a 3557-atom planetary gear, shown in side, end, and exploded views in Figure 2.

The entire assembly has twelve moving parts and is 4.3 nm in diameter and 4.4 nm in length, with a molecular weight of 51,009.844 daltons and a molecular volume of 33.458 nm^3. An animation of the computer simulation shows the central shaft rotating rapidly and the peripheral output shaft rotating slowly.

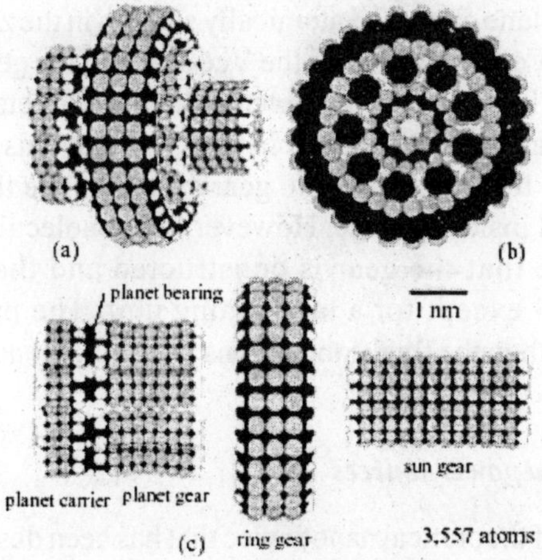

Figure. 2. End-, side-, and exploded-view of a 3557-atom planetary gear.

The small planetary gears, rotating around the central shaft, are surrounded by a ring gear that holds the planets in place and ensures that all components move in proper fashion. The ring gear is a strained silicon shell with sulphur atom termination; the sun gear is a structure related to an oxygen-terminated diamond (100) surface; the planet gears resemble multiple hexasterane structures with oxygen rather than CH bridges between the parallel rings; and the planet carrier is adapted from a Lomer dislocationarray created by R.Merkle and L.Balasubramaniam, linked to the planet gears using C C bonded bearings.

A rotational impulse dynamics study of this first generation planetary gear Goddard and colleagues found that at the normal operational rotation rates for which this component was designed, the gear worked as intended and did not overheat. Started from room temperature, the gear took a few cycles to engage, then rotated thermally stably at ~400 K. Only when the gear was severely overdriven to ~100 GHz did significant instabilities appear, although the device still did not self-destruct. One run at ~80 GHz showed excess kinetic energy causing gear temperature to oscillate up to 450 K above baseline.

Commenting on the ongoing design effort, Goddard suggested that an optimal configuration could have the functionality of a planetary gear but might have an appearance completely different from the macroscopic system, and offered an example: "Because a gear tooth in the xy plane cannot be atomically smooth in the z-direction, we may develop a Vee design so that the Vee shape of the gear tooth in the z-direction nestles within a Vee notch in the race to retain stability in the z-direction as the teeth contact in the xy plane. This design would make no sense for a macroscopic gear system since the gear could never be placed inside the race. However, for a molecular system one could imagine that the gear is constructed and that the race is constructed all except for a last joining unit. The parts could be assembled and then the final connections on the face made to complete the design."

Nanomotors and power sources

Another class of theoretical nanodevice that has been designed is a gas-powered molecular motor or pump. The device could serve either as a pump for neon gas atoms or (if run backwards) as a motor to convert neon gas pressure into rotary power. The helical rotor has a grooved cylindrical bearing surface at each end, supporting a screwthreaded cylindrical segment in the middle. In operation, rotation of the shaft moves a helical groove past longitudinal grooves inside the pump housing.

There is room enough for small gas molecules only where facing grooves cross, and these crossing points move from one side to the other as the shaft turns, moving the neon atoms along.

Goddardreported that preliminary molecular dynamics simulations of the device showed that it could indeed function as a pump, although it is not very energy-efficient so further refinement of this initial design is warranted.

Almost all such design research in molecular nanotechnology is restricted to theory and computer simulation, which allows the design and testing of large structures or complete nanomachines and the compilation of growing libraries of molecular designs.

Future nanosystem simulations may require 1-100 million atoms to be considered explicitly, demanding further improvements in present-day molecular dynamics methodologies which have only relatively recently entered the multi-million atom range. On the experimental pathway, Montemagno and Bachand modified a natural biomotor to incorporate nonbiological parts, creating the first artificial hybrid nanomotor.

Using the tools of genetic engineering, they added metal-binding amino acid residues to ATPase, a ubiquitous enzyme whose moving part is a central protein shaft that rotates in response to electrochemical reactions with each of the molecule's three proton channels. Each motor molecule bonded tightly to nanoscale nickel pedestals prepared by electron beam lithography.

Properly oriented motor molecules 12 nanometers in diameter were then attached to the pedestals with a precision approaching 15 nanometers, and a silicon nitride bar a hundred nanometers long was bound to the rotor subunit of each motor molecule, all by self-assembly. In a microscopic video presentation, dozens of bars could be seen spinning like a field of tiny propellers.

The group's first integrated molecular motor ran for 40 minutes at 3-4 revolutions per second, but subsequent motors have been operated for hours continuously by providing a surplus of ATP. Montemagno is also trying to build a solar-powered, biomolecular motor-driven autonomous nanodevice, wherein light energy is converted into ATP which then serves as a fuel source for the motor, and also a chemical means of switching his hybrid motors on and off reliably: By engineering a secondary binding site tailored to a cell's signalling cascade, he plans to use the sensory system of the living cell to control nanodevices implanted within the cell.

Montemagno envisions tiny chemical factories operating inside living cells. He speculates that these nanofactories could be targeted to specific cells, such as those of tumours, where they would synthesize and deliver chemotherapy agents.

Also following the bio-nano pathway is Mavroidis's group which in late 2003 received a $1 M four-year NSF grant to produce a viral protein linear nanomotor prototype by 2007, that will "pave the way for development of complete nanorobotic assemblies" and later "make up the systems that travel the bloodstream or perform other unprecedented tasks in medicine and industry." Others have operated chemically-powered DNA-based nanomotors.

Other experimental nanomotor research includes the 78-atom chemically-powered rotating motor synthesized in 1999 by Kelly, 139 a chemically-powered rotaxane-based linear motor exerting ~100 pN of force with a 1.9 nm throw and a ~250 sec contraction cycle by Stoddart's group, a UV-driven catenane-based ring motor by Wong and Leigh, and an artificial 58-atom motor molecule that spins when illuminated by solar energy by Feringa.

Also in 2003, the Zettl group created an electrically powered 550-nm wide nanomotor by first depositing a number of multiwalled nanotubes on the flat silicon oxide surface of a silicon wafer, then using electron beam lithography to simultaneously pattern a 110-300 nm gold rotor, nanotube anchors, and opposing stators around the chosen nanotubes, then annealing the rotor to the nanotubes, after which the surface was selectively etched to provide sufficient clearance for the rotor.

When the stators were alternately charged with 50 volts of direct current, the gold rotor rocked back and forth up to 20 degrees, making a torsional oscillator. A strong electrical jolt to the stators jerked the rotor and broke the outer wall of the nested nanotubes, allowing the outer nanotube and attached rotor to spin freely around the inner nanotubes as a nearly frictionless bearing. The oscillating rotor might be used to generate microwave frequency oscillations possibly up to a few gigahertz, or the spinning rotor could be used to mix liquids in micro fluidic devices.

Digital nanocomputers

Truly effective medical nanorobots may require onboard computers to allow a physician to properly monitor and control their work. In 2000, a collaborative effort between UCLA and Hewlett Packard produced the first laboratory demonstration of completely reversible room temperature molecular switches that could be employed in nanoscale memories, using mechanically interlinked ring molecules called catenanes, and there has been much recent progress with nanotube- and nanorod-based molecular electronics.

Several private companies are pursuing the first commercial molecular electronic devices including memories and other computational components of nanocomputers using techniques of self-assembly, and there is also the possibility of low-speed biology-based digital nanocomputers.

Self-Assembly and Directed Parts Assembly

Self-assembly of mechanical parts

Perhaps the best-known self-assembling molecular systems include those which form ordered monomolecular structures by the coordination of molecules to surfaces, called self-assembled monolayers (SAMs), self-assembling thin films or Langmuir-Blodgett films, self-assembling lipidic micelles and vesicles, and selforganising nanostructures.

In many of these systems, a single layer of molecules affixed to a surface allows both thickness and composition in the vertical axis to be adjusted to 0.1-nm by controlling the structure of the molecules comprising the monolayer, although control of in-plane dimensions to <100 nm is relatively difficult.

Several attempts have been made to achieve self-assembly of small mechanical parts to avoid direct parts grasping, and Saitougives a simple example of "sequential random bin picking" in which a process of sequential mating of a random pair of parts drawn from a parts bin which initially contains a random assortment of parts can produce the mating of a desired pair of parts. Griffith suggests expanding the toolbox of self-assembly by including dynamic components that emulate enzymatic allostery, and presents a simple "mechanical enzyme" analog-a 2-bit mechanical state machine that programmatically self-assembles while floating at an interface between water and poly-.uorodecalin.

The mechanical state machine has a mechanical flexure that acts as the 'switch' in the state machine, making a mechanical allosteric enzyme. As more component types are added the challenge is to avoid any undesirable local energy minima- necessitating the development of energy vs florientation modelling tools. The programming of engineered sequences of such conformational switches can allow the self-assembly of quite complicated mechanical structures.

Saitou has presented a model of self-assembling systems in which assembly instructions are written as conformational switches-local rules that specify conformational changes of a component. The model is a self-assembling automaton explicitly inspired by the Penrose self-replicating blocks and by Hosokawa's self-assembling triangular parts with embedded switches. Saitou claimshis model of self-assembling automata can also be applied to self-assembly in 2- or 3-dimensions.

Guidedor directed self-assembly has become a growing research area. Yeh and Smith have described a process of fluidic self-assembly of optoelectronic devices, Rothemund and Winfreehave described a tile assembly model for pseudocrystalline self-assembly, Breivikhas designed and patented a set of self-replicating physical polymers, and Gracias et al. have impressed electrical circuits including LEDs on the surfaces of copperpolyimide truncated octahedra each ~1 mm in diameter, then induced these octahedra to self-assemble into specified 3-D electrical networks of up to 12 devices by co-melting of opposing solder spots.

Whitesides et al. first demonstrated capillary force driven assembly of a simple circuit and other structures from millimetre-scale components, and electrostatic self-assembly, and some of this work has since been extended to the fluidic self-assembly of microscale parts, including "micro-origami" or "silicon origami", as well as mesoscopic nucleic acid analogs. The dynamics of Brownian self-assembly, the theory of designable self-assembling molecular machine structures, and the computational modelling of self-assembly processes are beginning to be addressed.

DNA-directed assembly

Early mechanical nanorobots might be assembled, at least in part, from DNA. The idea of using DNA to build nanoscale objects has been pioneered by Nadrian Seeman at New York University. Two decades ago, Seeman recognised that a strand of DNA has many advantages as a construction material. First, it is a relatively stiff polymer. Its intermolecular interaction with other strands can be readily predicted and programmed due to the base-pair complementarity of nucleotides, the fundamental building blocks of genetic material. DNA also tends to self-assemble. Arbitrary sequences are readily manufactured using conventional biotechnological techniques, and DNA is readily manipulated and modified by a large number of enzymes.

During the 1980s, Seeman worked to develop strands of DNA that would zip themselves up into more and more complex shapes-first tiny squares, then three-dimensional stick-figure cubes comprised of 480 nucleotides each, then a truncated octahedron containing 2550 nucleotides. By the mid-1990s, Seeman could fabricate nanoscale DNA stick figures of almost any regular geometric shape, by the billions per batch. In 1999, Seeman reported the construction of a mechanical DNA-based device that might serve as the basis for a nanoscale robotic actuator.

The mechanism has two rigid double-stranded DNA arms a few nanometers long that can be made to rotate between fixed positions by introducing a positively charged cobalt compound into the solution surrounding the molecules, causing the bridge region to be converted from the normal B-DNA structure to the unusual Z-DNA structure. The free ends of the arms shift position by 2-6 nanometers during this fully reversible structural conversion, like a hinge opening and closing. A large version of the device might function as an elbow, while smaller devices could serve as finger joints. By 2002, Seeman's group had demonstrated a mechanical DNA-based rotary motor and reported the design and construction of 2-dimensional DNA arrays which might serve as templates for nanomechanical assembly.

Seeman is now collaborating with genetic engineers and computational chemists to achieve "the design and fabrication of practical nanoscale devices" and "to make rapid progress in demonstrating DNA based nanoscale devices," including "sequence-dependent devices [that] can provide the diversity of structures necessary for nanorobotics."

In other labs: sequence-specific DNA hybridization is used to bend silicon microcantilevers, Alberti and Mergny synthesized a sequence-dependent DNA "piston" composed of a 21-base oligonucleotide that displays a 5-nm two-stroke linear motor type movement, Li and Tanhave made a single-DNA-molecule inchworm motor, and Shu and Guosynthesized a 30-nm long chimeric pRNA (DNA-packaging) motor made from six strands of RNA surrounding a center strand of DNA-in the presence of ATP, the RNA strands push the DNA axle in succession, spinning it around producing 50-60 pN of force. Yurke and Turber field synthesized another DNA actuator using three single strands of artificial DNA which, when placed together, find their complementary partners and self-assemble to form a V-shaped structure.

The open mouth of this nanotweezer can be made to close by adding a special "fuel" strand which binds to the single-stranded DNA dangling from the ends of the arms of the tweezers and zips them closed, moving from a ~7 nm separation to a ~1 nm separation in ~13 sec per cycle. A special "removal" strand, when added, binds to the fuel strand and pulls it away, opening the nanotweezers again. More recent workhas focused on a continuously running DNA nanomotor. Reif's group has devised X-shaped DNA tiles that link up in a square grid with some of the strands consisting of sections of DNA that can lengthen and shorten by 6.8 nm like tiny pistons, making a net whose mesh size can expand or contract under chemical control.

Protein-directed assembly

While most enzymes in cells are involved in manipulating small molecules <0.25 kD, there are several classes of enzymes involved in manufacturing complex covalently bound molecules such as vitamins, enzyme cofactors, antibiotics, and toxins with masses up to ~3 kD. Molecules even larger than this are manipulated by tRNA-synthetase (a 40-100 kD enzyme that manipulates ~30 kD tRNAs), the spliceosome, the ribosome, the proteosome, and the DNA replication complex.

Many of these protein manipulators employ "parts insertion" or "threading" maneuvers, such as the clamp and bridge helix mechanisms in RNA polymerase II that act as a translocation ratchet to feed DNA through the enzyme interior in order to produce mRNA. By designing synthetic enzymes which might be called "nanopart synthetases," possibly using synthetic amino acids, we can envision grabbing molecular parts in a solution and then, as the enzyme folds, bringing these parts into proper alignment and causing them to react, exemplifying protein-directed parts assembly.

Ratchet-action protein-based molecular motors are wellknown in biology, conformational cascades of a special genetic variant of yeast cell prions have already been used to assemble silver- and gold-particle-based nanowires, and the GTPase dynamin mechano enzyme-which self-assembles into rings or spirals, wrapping around the necks of budding vesicles and squeezing, pinching them off, during cellular endocytosis-is wellknown.

Smithhas used methyl transferase-directed addressing of fusion proteins to DNA scaffolds to construct a molecular camshaft as a exemplar protein/nucleic acid biostructure. Protein-protein binding specificity can bend silicon microcantilevers. Genetically engineered chaperonin protein templates (chaperone molecules) can direct the assembly of gold nanoparticles (1.4, 5, or 10 nm) and CdSe semiconductor quantum dots (4.5 nm) into nanoscale arrays.

Microbe- and Virus-Directed Assembly

Artificial microbes might also be employed in molecular parts fabrication. One strain of bacteria is known to fabricate single crystals of pure silver in specific geometric shapes such as equilateral triangles and hexagons, up to 200 nm in size, and microorganisms can accumulate materials and synthesize inorganic structures composed of bismuth, CdS, gold, magnetite, silica, and silver. As for microbe-directed parts assembly, Kondo et al. used a grooved film to train the bacterium Acetobacter xylinum to exude neat ribbons of cellulose along the prepared track at a rate of 4 microns/minute.

Fibroblasts can be genetically engineered, are capable of crosslinking collagen fibers (a "covalent parts joining" type of operation), and can apply ~100 pN forces while embedded in a 3-

dimensional collagen lattice. Although ECM (extracellular matrix) strand positioning is stochastic in natural broblasts, cell functionality and ECM network characteristics can be altered by chemotactic factors, contact guidance and orientation, hypoxia, and local mechanical stress.

To establish digital control over microorganisms, genetic circuits that can function as switchesor computational logic elements such as AND, NAND, and NOR gates are under active investigation. In 2000 Gardner et al. added a memory device to an E. coli bacterium using two inverters for which the output protein of each is the input protein of the other. Elo witz and Leiblermade an oscillator with three inverters connected in a loop-in one test of their system, "a fluorescent protein became active whenever one of the proteins was in its low state...the result was a population of gently twinkling cells like flashing holiday lights." By 2002, Weisshad created a five-gene circuit in E. coli that could detect a specific chemical in its surroundings and turn on a fluorescent protein when the chemical concentration falls within preselected bounds. Bacteria can serve as physical system components.

Tung et al. are attempting to incorporate living bacteria into microelectro mechanical systems (MEMS) devices to form living cell motors for pumps, valves, and conveyor belts. Turner and colleagues have affixed a film of Serratia marcescens bacteria onto tiny beads, allowing the microbes' rotating appendages to carry the beads along, then applied the film inside tiny tubes, whereupon the gyrating bacterial arms blend fluids twice as fast as diffusion alone. Montemagno has combined living cells with isolated MEMS structures to create cell-powered mechanical motors. In one experiment in 2003, a lithographically produced U-shaped structure 230 microns wide is attached to a cardiac muscle cell like a tiny prosthesis.

When presented with glucose solution, the muscle cell contracts repeatedly, causing the mechanical structure to "walk" at a speed of ~46 microns/min with a repetition rate controlled by the spring constant of the MEMS structure. Viral shells also provide useful templates for nanoscale assembly. Belcher employs virus capsid shells as scaffolds for the directed nanoassembly of nanoparticles such as quantum dots, in a process of "biomimetic synthesis of nonbiological inorganic phases with novel electronic and magnetic properties directed by proteins and synthetic analogs."

In one experiment, a genetically engineered M13 bacteriophage with a specific recognition moiety for zinc sulphide nanocrystals assembles a ZnS-containing film having nanoscale ordering and 72-micron-sized domains. Engineered viral coat proteins can be used as scaffolds for nanomaterials synthesis and self-assembly, including self-assembled monolayers.

Positional Assembly

As machine structures become more complex, getting all the parts to spontaneously self-assemble in the right sequence is increasingly difficult. To build complex nonperiodic structures, it makes more sense to design a mechanism that can assemble a molecular structure by what is called positional assembly-that is, picking and placing molecular parts. A device capable of positional assembly would work much like the robot arms that manufacture cars on automobile assembly lines. In this approach, the robot manipulator picks up a part, moves it to the workpiece, installs it, then repeats the procedure over and over with many different parts until the final product is fully assembled.

One of the leading proponents of positional assembly at the molecular scale is Zyvex Corp, the first engineering firm to espouse an explicit goal of using positional assembly to manufacture atomically precise structures, or more specifically, "a user-controlled fabrication tool capable of creating molecularly precise structures with 3-dimensional capability in an economically viable manner."

Zyvex has already demonstrated the ability to positionally assemble large numbers of MEMSscale parts, and has demonstrated the ability to use three independently-controlled inch-long robotic arms to manipulate tiny carbon nanotubes in three dimensions, under the watchful eye of a scanning electron microscope that can monitor objects and motions as small as 6 nanometers at near-video scan rates.

Agilent Laboratories has created an ultra-high-precision micromover platformcapable of providing linear two-dimensional movement in steps of 1.5 nanometers, the width of about 9 bonded carbon atoms. The core of the micromover is a stepper actuator or linear motor that does not rotate, but instead steps right to left or front to back.

The platform can travel a total of 30 micrometers in each direction in 2.5 milliseconds; since each micrometer is made up of 1,000 nanometers, the micromover would take approximately 20,000 steps to traverse 30 micrometers, a distance which is about half the width of a single human hair. Martel's group at MIT has worked on a similar nano-positioning device called the NanoWalker. Kim and Liebercreated the first general-purpose nanotweezer whose working end is a pair of electrically controlled carbon nanotubes made from a bundle of multiwalled carbon nanotubes.

To operate the tweezers, a voltage is applied across the electrodes, causing one nanotube arm to develop a positive electrostatic charge and the other to develop a negative charge. Kim and Lieber have successfully grasped organelle-sized 500-nanometer clusters of polystyrene spheres, and have removed a semiconductor wire 20 nanometers wide from a mass of entangled wires, using tweezer arms about 50 nanometers wide and 4 microns long, but the technique creates a large electric field at the tweezer tips which can alter the objects being manipulated.

In 2001, Boggild's groupused standard micromachining processes to carve from a tiny slab of silicon an array of cantilevered micro-pliers which could be opened and closed electrically. Boggild then used an electron beam to grow a tiny carbon nanotweezer arm from the end of each cantilever, angled so that the tips were only 25 nanometers apart, making a better-controlled nanotweezer.

Nanotube-based nanotweezers have since been reported by others. Precise positional covalent attachment of molecules to surfaces is also being pursued. Blackledge et al. used a palladium-coated SFM (Scanning Force Microscope) tip to chemically modify terminal functional groups on an organosiloxane-coated surface to create biotin-streptavidin assemblies in patterns with minimum 33 nm line widths.

Diaz et al. employ redox probe microscopy (RPM) in which an SFM tip is modified with redox-active materials, whereupon the interactions between tip and an adsorbate or between tip and a surface are modulated by the electrode potential. This system has also been used as a microtweezer to manipulate and position objects. Hla and Riederhave reviewed recent progress in using scanning tunnelling microscopy (STM) to manipulate and synthesize individual molecules.

The ultimate goal of molecular nanotechnology is to develop a manufacturing technology that can inexpensively manufacture most arrangements of atoms that can be specified in molecular detail-including complex arrangements involving millions or billions of atoms per product object, as in the hypothesized medical nanorobots. This will provide the ultimate manufacturing technology in terms of its precision, flexibility, and low cost. Two central mechanisms have been proposed to achieve these goals at the molecular scale: programmable positional assembly including fabrication of diamondoid structures using molecular feedstock, and massive parallelism of all fabrication and assembly processes.

Diamond mechanosynthesis

There is widespread interest in the exceptional properties of diamond-it has extreme hardness and strength, high thermal conductivity, low frictional coefficient, chemical inertness, and a wide bandgap. Recent investigations have been driven by the many emerging applications for diamond in MEMS mechanical and electromechanical devices, optics, radiology, biochemical synthesis and medicine, but most especially in various electronics devices. A method for the precise manufacture of microscopic and nanoscale diamond structures would have tremendous utility in science and industry.

In contrast to high-pressure diamond synthesis and low-pressure gas-phase diamond synthesis of diamond via chemical vapour deposition or CVD, positional mechanosynthesis has been proposed by Drexler for the precise manufacture of diamond structures. Mechanosynthesis aims to achieve site-specific chemical synthesis by inducing chemical transformations controlled by positional systems operating with atomic-scale precision, thus enabling direct positional selection of reaction sites on the workpiece. The scanning tunnelling electron microscope (STM) has demonstrated an ability to manipulate surface structures atom by atom, and many proposed methods involve the use of an SPM to direct chemical reactions on the surface by:

(1) delivering an electric field to a sub-nanometer region of a surface to activate a chemical reaction,

(2) manipulating the chemistry of the tip to make it act as a catalyst which can then be introduced precisely into the region of desired reaction, or

(3) delivering mechanical energy from the tip to activate surface reactions (mechanosynthesis).

The reaction selectivity of all these methods relies on the exponential dependence of reaction rates on the activation barrier, which is lowered for surface reactions in a precisely defined area of the surface during mechanosynthesis.

The principal challenge in diamond mechanosynthesis is the controlled addition of carbon atoms to the growth surface of the diamond crystal lattice. The theoretical analysis of carbon atom (or dimer) placement on diamond has involved many researchers including Cagin, Drexler, Dzegilenko, Freitas, Goddard, Mann, Merkle, Peng, Saini, Srivastava, and Walch.

The feasibility of precisely inserting individual carbon atoms, small hydrocarbon species, or small clusters of carbon atoms on a C(111) or C(100) diamond surface at specific sites was initially supported first by the computational work of Walch and Merkle. Walch and Merkle analysed several mechanosynthetic reactions, including placement of carbon dimer onto a C(111) surface, insertion of a positionally controlled carbene into that dimer, and the insertion of a positionally controlled carbene into a surface dimer on a C(100) surface using a 9-atom cluster to model the diamond surface.

The latter insertion can take place with no barrier (according to computational results based on ab initio calculations using Gaussian with a 6-31 G basis set and B3LYP density functional) provided the approach trajectory is appropriate. Subsequent removal of the mechanosynthetic tool tip using an appropriate withdrawal trajectory is predicted to leave a single carbon atom in the bridged position on the dimer.

Classical molecular dynamics simulations by Dzegilenko et al. showed that a single weaklybonded carbon dimer could be selectively removed from the upper terrace of a reconstructed diamond C (100) -

(2×1) surface by a carbon nanotube tip chemically modified with a C_2 carbene radical species strongly bonded to the end cap of the tip.

When planar C_6H_2 (methenylidene cyclopentene) is brought up to the C(100) surface, either a C_3H moiety with two lower C atoms of the tip initially deposited onto the four-fold locations forming bonds with C atoms of two neighbouring surface dimmers is attached, or else a C_4H_2 fragment is adsorbed atop two C atoms of neighbouring surface dimmers, with the reaction outcome depending critically upon the initial tip-surface distances, the tip trajectory, and various allowed but undesired tip re-arrangements.

In 2003, Merkle and Freitasproposed a new family of mechanosynthetic tools intended to be employed for the placement of two carbon atoms-a CC dimer- onto a growing diamond surface at a specific site. Their analysis used density functional theory with Gaussian 98 to focus on specific group IV-substituted biadamantane tooltip structures and evaluate their stability and the strength of the bond they make to the CC dimer.

Considering a dimer bonded to two group IV supporting atoms (silicon, germanium, tin, or lead), this series of elements forms progressively weaker bonds to carbon, so the proposed tooltips will likewise be progressively more weakly bound to the carbon-carbon (CC) dimer. The supporting group IV atoms are part of two substituted adamantane ($C_{10}H_{16}$) frameworks that position and orient them.

The tooltip molecule, a bi-silaadamantane dicarbon, is only the apex of a complete tool. In a full mechanosynthetic apparatus, a somewhat larger version of this molecule would likely be required so that the active tip could be held and positioned via a rigid handle structure. Initial and subsequent ab initio molecular dynamics simulations of these tool tips has confirmed the successful operation of the Ge tooltip at room temperatures.

Although pick-and-place of individual carbon atoms or carbon dimmers has not yet been demonstrated experimentally using scanning probe microscope tips, in 1985 Becker and Golovchenckoused voltage pulses on an STM tip to extract a single germanium atom from the (111) surface of a sample. STMs have been used to bind silicon atoms to the tip, first pulling the atoms off the surface of a Si(111) crystal face and then re-inserting them back into the crystal, and segments of individual dimer rows of silicon atoms have been extracted from the Si(100) face to create structures with atomically straight edges and lateral features that are only one dimer in width.

Ho and Leehave demonstrated the first repeatable site specific mechanosynthetic covalent bonding operation of a diatomic carbon-containing molecule, $Fe(CO)_2$, on a crystal surface, albeit electrically-mediated. Most recently, a near contact atomic force microscope operated at low temperature has been used for the vertical manipulation of selected single silicon atoms from the Si(111)-(7 ×7) surface, the first experimental demonstration of pure mechanosynthesis involving the solely mechanical removal of a selected silicon atom from its equilibrium position at the surface without otherwise perturbing the (7×7) unit cell, as well as the mechanosynthetic deposition of a single atom on a created surface vacancy.

These results, both theoretical and experimental, support the general feasibility of molecular positional operations that can modify a diamond workpiece, adding or removing small hydrocarbon clusters on that workpiece or even adding and removing single atoms or dimmers under appropriate vacuum conditions.

Repeated application of these basic operations should allow building up complex and atomically precise molecular structures, permitting the manufacture of a wide range of nanoscale diamond structures with atomically precise features.

Massively parallel manufacturing

Massively parallel assembly is the key to the economic viability of molecular manufacturing. Biology provides perhaps the best example of the power of massive parallelism in assembly, such as polysomes in living cells. The difference between serial and parallel processing is similarly crucial in molecular manufacturing, where the basic parts are very small.

If a typical molecularly precise simple component is 1 nm^3 in volume, then to manufacture a 1 cm^3 volume of molecularly precise product requires the assembly of 1000 billion billion (10^{21}) individual simple molecular components-even at a 1 GHz operating frequency, serial atom-by-atom manufacturing of a single object would take many thousands of years, clearly not economically viable. But with parallel manufacti ring, vast numbers of molecular components can be processed simultaneously, reducing batch processing times to days, hours, or even less. At least two such techniques for performing massively parallel positional assembly have been identified:

(1) massively parallel manipulator arrays and

(2) self-replicating systems.

Massively parallel manipulator arrays would use a very large array of independently actuated manipulation devices to process a very large number of molecular precise components simultaneously to build a larger product object. In order to fabricate large numbers of nanoparts and nanoassemblies, massively parallel scanning probe microscopes (SPM) arrays and microscale SPMs would be most convenient.

Force-sensing devices such as piezoelectric, piezoresistive and capacitive microcantilevers make it possible to construct microscale AFMs on chips without an external deflection sensor. For example, in 1995 Itoh and colleagues fabricated an experimental piezoelectric ZnO_2-on-SiO_2 microcantilever array of ten tips on a single silicon chip.

Each cantilever tip lay ~70 microns from its neighbour, and measured 150 microns long, 50 microns wide and 3.5 microns thick, or ~26,000 $micron^3$/device, and each of the devices could be operated independently in the z-axis up to near their mechanical resonance frequencies of 145-147 KHz at an actuation sensitivity of ~20 nm/volt-for instance, 0.3-nm resolution at 125 KHz.

Parallel probe scanning and lithography was been achieved by Quate's group, progressing from simple piezoresistive microcantilever arrays with 5 tips spaced 100 microns apart and 0.04-nm resolution at 1 KHz but only one z-axis actuator for the whole array, to arrays with integrated sensors and actuators that allow parallel imaging and lithography with feedback and independent control of each of up to 16 tips, with scanning speeds up to 3 mm/sec using a piezoresistive sensor, and by 1998 to arrays of 50-100 independently controllable AFM probe tips mounted in 2-D patterns with 60 KHz resonances, including a 10 ×10 cantilevered tip array fabricated in closely spaced rows using throughwafer interconnects on a single chip.

Similarly, by 1997 MacDonald's grouphad built and tested an array of micro-STMs on the surface of an ordinary silicon chip, with each tip on a cantilever 150 microns long with 3-D sensing and control-the largest prototype array had 144 probes, arranged in a square consisting of 12 rows of 12 probes each, with individual probe needles about 200 microns apart.

Researchers in the Millipede project at IBM's Zurich Research Laboratory used conventional microlithography to fabricate scanning probe tip arrays of up to 1024 individual tips to achieve terabit-per-square-inch data storage densities, Mirkin's group constructed an array of 10,000 microscope tips with each capable of acting independently from the others, and at least one "electronic nose" microcantilever array has been fabricated with millions of interdigitated cantilevers on a single chip.

Yet another alternative is Zyvex's patented Rotapod ™ exponential assembly design concept, in which a single robotic arm on a wafer makes a second robotic arm on a facing surface by picking up micron-size lithographically produced parts-carefully laid out in advance in exactly the right locations so the tiny robotic arm can find them- and assembling them.

The two robotic arms then make two more robotic arms, one on each of the two facing surfaces. These four robotic arms, two on each surface, then make four more robotic arms. This process continues with the number of robotic arms steadily increasing in the pattern 1, 2, 4, 8, 16, 32, 64, etc., until some manufacturing limit is reached.

Thus a single manipulator uses supplied parts to build a large manipulator array which can subsequently undertake the desired massively parallel manufacturing operations. In 2001, Zyvex was awarded a $25 million, five-year, National Institute of Standards and Technology (NIST) Advanced Technology Program government contract to develop prototype microscale assemblers using microelectro mechanical systems (MEMS) and nanoelectro mechanical systems (NEMS) for prototype nanoscale assemblers.

Self-replicating systems would achieve massively parallel assembly first by fabricating copies of themselves, then allowing those copies to fabricate further copies, resulting in a rapid increase in the total number of systems. Once the population of replicated manipulator systems was deemed large enough, the manipulator population would be redirected to produce useful product objects, rather than more copies of itself. Self-replicating systems are widely found in natural biological systems but have not been pursued explicitly in macroscale manufacturing for at least two reasons:

(1) the widespread but erroneous perception of great technical difficulty, and

(2) the correct perception that such massive parallelism is unnecessary for traditional macroscale manufacturing.

Nevertheless, ever since John von Neumann's theoretical studies of replicating systems in the 1940s and 1950s, and the well-known 1980 NASA engineering study of self-replicating lunar factories, manufacturing automation has been slowly progressing toward the goal of the fully self-replicating factory-including most notably Fujitsu Fanuc's nearly "unmanned" robot factory in Yamanashi Prefecture that uses robot arms to make robot arms. It is worth noting that self-replicating systems can be fully remote-controlled, fully autonomous, or various combinations in between.

In the last few years there has been renewed research interest in the challenge of mechanical self-replicating systems, in part due to the realisation that replication can be a fundamentally simple process. Today there are several ongoing university research programmes, both theoretical and experimental, on mechanical (nonbiological) self-replicating machines. The biotechnology and molecular engineering communities are just beginning to seriously study mechanical replicators operating in the nanoscale size domain.

Note that for the foreseeable future it is likely that onboard storage of information will not be required by nanomechanical replicators. One example of an inherently safe and flexible approach is the broadcast architecture, wherein control information is broadcast by any of several means to the replicating component.

The physical replicator becomes, in essence, a remote-controlled manipulator receiving instructions from the outside that guide it, step by step, in assembling a second remote-controlled manipulator. After some number of repeat cycles, the result is a large number of identical remote-controlled manipulators. These manipulators can then be used to assemble large numbers of useful product objects by altering the stream of instructions sent to the population of replicated manipulator devices. Conceptual system designs for molecular manufacturing are extensively reviewed by Freitas and Merkle.

Medical Nanorobot Designs

The idea of placing autonomous self-powered nanorobots inside of us might seem a bit odd, but actually the human body already teems with such nanodevices. For instance, more than 40 trillion single-celled microbes swim through the colon, outnumbering the tissue cells almost ten to one. Many bacteria move by whipping around a tiny tail, or flagellum, that is driven by a 30-nanometer biological ionic nanomotor powered by pH differences between the inside and the outside of the bacterial cell.

Our bodies also maintain a population of more than a trillion motile biological nanodevices called fibroblasts and white cells such as neutrophils and lymphocytes, each measuring perhaps 10 microns in size. These beneficial natural nanorobots are constantly crawling around inside of us, repairing damaged tissues, attacking invading microbes, and gathering up foreign particles and transporting them to various organs for disposal from the body.

There are ongoing attempts to build MEMS-based microrobots intended for in vivo use. For example, the "MR-Sub" project of the NanoRobotics Laboratory of Ecole Polytechnique in Montreal will use a Magnetic Resonance Imaging (MRI) system as a means of propulsion for a microrobot in the blood vessels. In this approach, a variable MRI magnetic field would generate a magnetic force on a robot containing ferromagnetic particles, providing a miniaturised system of propulsion able to develop sufficient power to direct a small device through the human body. Applications of the first generation prototype might include targeted drug release, the reopening of blocked arteries, or taking biopsies.

The project is currently gathering necessary information to define design rules for this type of microrobot, with a long-term goal "to further miniaturize the system and to create a robot made up of nanometric parts," making it "possible to carry out medical applications in the blood vessels which are still inaccessible." Other approaches to MEMS-based microrobots intended for in vivo use have been described in the literature, including the magnetically-controlled "cytobots" and "karyobots" proposed by Chrusch et al. for performing wireless intracellular surgery.

There are preliminary proposals for hybrid bionanorobots that could be constructed using currently foreseeable technologies. For example, Montemagnoplans to use his modified ATPase motors to create a nanorobot that acts as a "pharmacy in a cell" by entering a cell, grabbing proteins produced by the cell that will not be used, and storing them until they are needed later by the patient.

The device would consist of a tiny nickel drum, attached to the ATP-powered biological motor, which is coated with antibodies that adsorb the target molecules, whereupon an electric field pulls the molecules to a storage chamber and holds them in place.

The greatest power of nanomedicine will emerge in a decade or two when we learn to design and construct complete artificial nanorobots using diamondoid nanometerscale parts and subsystems including sensors, motors, manipulators, power plants, and molecular computers.

There are many possibilities but the development pathway will be long and arduous. First, theoretical scaling studies are used to assess basic concept feasibility. These initial studies would then be followed by more detailed computational simulations of specific nanorobot components and assemblies, and ultimately full systems simulations, all thoroughly integrated with additional simulations of massively parallel manufacturing processes from start to finish consistent with a design-for-assembly engineering philosophy.

Once molecular manufacturing capabilities become available, experimental efforts may progress from component fabrication and testing, to component assembly, and finally to prototypes and mass manufacture, ultimately leading to clinical trials.

Respirocytes

The artificial mechanical red blood cell or "respirocyte" is a bloodborne spherical 1-micron diamondoid 1000- atmosphere pressure vessel (Figure 3) with active pumping powered by endogenous serum glucose, able to deliver 236 times more oxygen to the tissues per unit volume than natural red cells and to manage carbonic acidity.

The nanorobot is made of 18 billion atoms precisely arranged in a diamondoid pressure tank that can be pumped full of up to 3 billion oxygen (O_2) and carbon dioxide (CO_2) molecules. Later on, these gases can be released from the tank in a controlled manner using the same molecular pumps. Respiroc ytes mimic the action of the natural hemoglobin-filled red blood cells. Gas concentration sensors on the outside of each device let the nanorobot know when it is time to load O_2 and unload CO_2 (at the lungs), or vice versa.

Figure 3. An artificial red cell—the respirocyte. 285 Designer Robert A.Freitas, Jr.© 1999, Forrest Bishop. Used with permission.

An onboard nanocomputer and numerous chemical and pressure sensors enable complex device behaviours remotely reprogrammable by the physician via externally applied acoustic signals. Each respirocyte can store and transport 236 times as much gas per unit volume as a natural red cell. So the injection of a 5 cc therapeutic dose of 50% respirocyte saline suspension, a total of 5 trillion individual nanorobots, into the human bloodstream can exactly replace the gas carrying capacity of the patient's entire 5.4 liters of blood.

If up to 1 liter of respirocyte suspension could safely be added to the human bloodstream, this could keep a patient's tissues safely oxygenated for up to 4 hours in the event a heart attack caused the heart to stop beating, even in the absence of respiration. Primary medical applications of respirocytes will include transfusable blood substitution; partial treatment for anemia, perinatal/ neonatal and lung disorders; enhancement of cardiovascular/ neurovascular procedures, tumour therapies and diagnostics; prevention of asphyxia; artificial breathing; and a variety of sports, veterinary, battle field and other uses.

Microbivores

An arti.cial mechanical white cell of microscopic size, called a "microbivore," has as its primary function to destroy microbiological pathogens found in the human bloodstream using a digest and discharge protocol. The microbivore is an oblate spheroidal nanomedical device measuring 3.4 microns in diameter along its major axis and 2.0 microns in diameter along its minor axis, consisting of 610 billion precisely arranged structural atoms in a gross geometric volume of 12.1 $micron^3$ and a dry mass of 12.2 picograms.

The device may consume up to 200 pW of continuous power while completely digesting trapped microbes at a maximum throughput of 2 $micron^3$ of organic material per 30-second cycle, which is large enough to internalise a single microbe from virtually any major bacteremic species in a single gulp. The nanorobots would be ~80 times more efficient as phagocytic agents than macrophages in terms of volume/sec digested per unit volume of phagocytic agent, and would have far larger maximum lifetime capacity for phagocytosis than natural white blood cells.

Microbi vores would fully eliminate septicemic infections in minutes to hours, whereas natural phagocytic defences-even when aided by antibiotics- can often require weeks or months to achieve complete clearance of target bacteria from the bloodstream. Hence microbivores appear to be up to ~1000 times faster-acting than either unaided natural or antibiotic-assisted biological phagocytic defences, and able to extend the therapeutic competence of the physician to the entire range of potential bacterial threats, including locally dense infections.

During each cycle of nanorobot operation, the target bacterium is bound to the surface of the bloodborne microbivore like a fly on flypaper, via species-specific reversible binding sites. Telescoping robotic grapples emerge from silos in the device surface, establish secure anchorage to the microbe's plasma membrane, then transport the pathogen to the ingestion port at the front of the device where the pathogen cell is internalised into a 2 micron^3 morcellation chamber.

After sufficient mechanical mincing, the morcellated remains of the cell are pistoned into a separate 2 micron^3 digestion chamber where a preprogrammed sequence of 40 engineered enzymes are successively injected and extracted six times, progressively reducing the morcellate ultimately to monoresidue amino acids, mononucleotides, glycerol, free fatty acids and simple sugars.

These simple molecules are then harmlessly discharged back into the bloodstream through an exhaust port at the rear of the device, completing the 30-second digestion cycle. This "digest and discharge" protocolis conceptually similar to the internalisation and digestion process practiced by natural phagocytes, except that the artificial process should be much faster and cleaner. For example, it is well-known that macrophages release biologically active compounds during bacteriophagy, whereas well-designed microbivores need only release biologically inactive effluent.

In the first half of the 21st century, nanomedicine should eliminate virtually all common diseases of the 20th century, and virtually all medical pain and suffering as well. It is a bright future that lies ahead for nanomedicine, but we shall all have to work very long and very hard to make it come to pass.

References

Howard J., Hudspeth A.J., and Vale R.D., "Movement of microtubules by single kinesin molecules", *Nature* 342: 154–158, 1989.

Kinosita K., Jr., Yasuda R., Noji H., Adachi K., "A rotary molecular motor that can work at near 100% ef.ciency", *Philos. Trans.: Biol. Sci.* 355 (1396): 473–489, 2000.

LymnR.W. and Taylor E.W., "Mechanism of adenosine triphosphate hydrolysis by actomyosin", *Biochemistry* 10: 4617–4624, 1971.

5

Nanoparticles

In nanotechnology, a particle is defined as a small object that behaves as a whole unit in terms of its transport and properties. It is further classified according to size: In terms of diameter, fine particles cover a range between 100 and 2500 nanometers, while ultrafine particles, on the other hand, are sized between 1 and 100 nanometers. Similarly to ultrafine particles, nanoparticles are sized between 1 and 100 nanometers, though the size limitation can be restricted to two dimensions.

Nanoparticles may or may not exhibit size-related intensive properties that differ significantly from those observed in fine particles or bulk materials. Nanoclusters have at least one dimension between 1 and 10 nanometers and a narrow size distribution. Nanopowders are agglomerates of ultrafine particles, nanoparticles, or nanoclusters. Nanometer sized single crystals, or single-domain ultrafine particles, are often referred to as nanocrystals. The term NanoCrystal® is a registered trademark of Elan Pharma International (EPIL) used in relation to EPIL's proprietary milling process and nanoparticulate drug formulations.

Nanoparticle research is currently an area of intense scientific research, due to a wide variety of potential applications in biomedical, optical, and electronic fields. The National Nanotechnology Initiative has led to generous public funding for nanoparticle research in the United States.

Although generally nanoparticles are considered an invention of modern science, they actually have a very long history. Specifically, nanoparticles were used by artisans as far back as in the 9th century Mesopotamia for generating a glittering effect on the surface of pot. Even these days pottery from the Middle Ages and Renaissance often retain a distinct gold or copper coloured metallic glitter. This so called lustre is caused by a metallic film that was applied to the transparent surface of a glazing. The lustre can still be visible if the film has resisted atmospheric oxidation and other weathering.

The lustre originates within the film itself, which contains silver and copper nanoparticles, dispersed homogeneously in the glassy matrix of the ceramic glaze. These nanoparticles were created by the artisans by adding copper and silver salts and oxides together with vinegar, ochre, and clay, on the surface of previously-glazed pottery. The object was then placed to a kiln and heated to about 600°C in a reducing atmosphere.

In the heat the glaze would soften, causing the copper and silver ions to migrate into the outer layers of the glaze. There the reducing atmosphere reduced the ions back to metals, which then came together forming the nanoparticles that give the colour and optical effects. Lustre technique shows that craftsmen had a rather sophisticated empirical knowledge of materials. The technique originates in the islamic world. As Muslims were not allowed to use gold in artistic representations, they had to find a way to create a similar effect without using real gold. The solution they found was using lustre.

Michael Faraday provided the first description, in scientific terms, of the optical properties of nanometer-scale metals in his classic 1857 paper "Experimental relations of gold (and other metals) to light." Much of the modern day studies of these objects have been conducted at the ESRF laboratory. Several techniques were used to characterise the chemical and physical properties of these lustre, such as Rutherford Backscattering Spectrometry (RBS), optical absorption in the visible-ultraviolet region, electron microscopy (TEM and SEM),

Properties of Nanoparticles

Nanoparticles are of great scientific interest as they are effectively a bridge between bulk materials and atomic or molecular structures. A bulk material should have constant physical properties regardless of its size, but at the nano-scale this is often not the case. Size-dependent properties are observed such as quantum confinement in semiconductor particles, surface plasmon resonance in some metal particles and superparamagnetism in magnetic materials.

The properties of materials change as their size approaches the nanoscale and as the percentage of atoms at the surface of a material becomes significant. For bulk materials larger than one micrometre the percentage of atoms at the surface is minuscule relative to the total number of atoms of the material. The interesting and sometimes unexpected properties of nanoparticles are partly due to the aspects of the surface of the material dominating the properties in lieu of the bulk properties.

Nanoparticles exhibit a number of special properties relative to bulk material. For example, the bending of bulk copper occurs with movement of copper atoms/clusters at about the 50 nm scale. Copper nanoparticles smaller than 50 nm are considered super hard materials that do not exhibit the same malleability and ductility as bulk copper. The change in properties is not always desirable.

Ferroelectric materials smaller than 10 nm can switch their magnetisation direction using room temperature thermal energy, thus making them useless for memory storage. Suspensions of nanoparticles are possible because the interaction of the particle surface with the solvent is strong enough to overcome differences in density, which usually result in a material either sinking or floating in a liquid. Nanoparticles often have unexpected visible properties because they are small enough to confine their electrons and produce quantum effects. For example gold nanoparticles appear deep red to black in solution.

Nanoparticles have a very high surface area to volume ratio. This provides a tremendous driving force for diffusion, especially at elevated temperatures. Sintering can take place at lower temperatures, over shorter time scales than for larger particles. This theoretically does not affect the density of the final product, though flow difficulties and the tendency of nanoparticles to agglomerate complicates matters. The large surface area to volume ratio also reduces the incipient melting temperature of nanoparticles.

Moreover nanoparticles have been found to impart some extra properties to various day to day products. Like the presence of titanium dioxide nanoparticles impart what we call as the self-cleaning effect, and the size being nanorange, the particles can't be seen. Nano Zinc Oxide particles have been found to have superior UV blocking properties compared to its bulk substitute. This is one of the reasons why it is often used in the sunscreen lotions.

Clay nanoparticles when incorporated into polymer matrices increase re-inforcement, leading to stronger plastics, verified by a higher glass transition temperature and other mechanical property tests. These nanoparticles are hard, and impart their properties to the polymer (plastic). Nanoparticles have also been attached to textile fibers in order to create smart and functional clothing.

NANOPARTICLE CLASSIFICATION

At the small end of the size range, nanoparticles are often referred to as clusters. Nanospheres, nanorods, and nanocups are just a few of the shapes that have been grown. Metal, dielectric, and semiconductor nanoparticles have been formed, as well as hybrid structures. Nanoparticles made of semiconducting material may also be labelled quantum dots if they are small enough (typically sub 10 nm) that quantisation of electronic energy levels occurs.

Such nanoscale particles are used in biomedical applications as drug carriers or imaging agents. Semi-solid and soft nanoparticles have been manufactured. A prototype nanoparticle of semi-solid nature is the liposome. Various types of liposome nanoparticles are currently used clinically as delivery systems for anticancer drugs and vaccines.

CHARACTERISATION

Nanoparticle characterisation is necessary to establish understanding and control of nanoparticle synthesis and applications. Characterisation is done by using a variety of different techniques, mainly drawn from materials science. Common techniques are electron microscopy [TEM,SEM], atomic force microscopy [AFM], dynamic light scattering [DLS], x-ray photoelectron spectroscopy [XPS], powder x-ray diffractometry [XRD], Fourier transform infrared spectroscopy [FTIR], Matrix-Assisted Laser-Desorption

Time-of-flight mass spectrometry [MALDI-TOF], and Ultraviolet-visible spectroscopy.

Whilst the theory has been known for over a century, the technology for Nanoparticle tracking analysis (NTA) allows direct tracking of the Brownian motion and this method therefore allows the sizing of individual nanoparticles in solution.

Fabrication

There are several methods for creating nanoparticles; attrition and pyrolysis are common methods. In attrition, macro or micro scale particles are ground in a ball mill, a planetary ball mill, or other size reducing mechanism. The resulting particles are air classified to recover nanoparticles. In pyrolysis, a vaporous precursor (liquid or gas) is forced through an orifice at high pressure and burned. The resulting solid (a version of soot) is air classified to recover oxide particles from by-product gases. Pyrolysis often results in aggregates and agglomerates rather than singleton primary particles.

A thermal plasma can also deliver the energy necessary to cause evaporation of small micrometre size particles. The thermal plasma temperatures are in the order of 10000 K, so that solid powder easily evaporates. Nanoparticles are formed upon cooling while exiting the plasma region. The main types of the thermal plasmas torches used to produce nanoparticles are dc plasma jet, dc arc plasma and radio frequency (RF) induction plasmas.

In the arc plasma reactors, the energy necessary for evaporation and reaction is provided by an electric arc which forms between the anode and the cathode. For example, silica sand can be vaporized with an arc plasma at atmospheric pressure. The resulting mixture of plasma gas and silica vapour can be rapidly cooled by quenching with oxygen, thus ensuring the quality of the fumed silica produced. In RF induction plasma torches, energy coupling to the plasma is accomplished through the electromagnetic field generated by the induction coil.

The plasma gas does not come in contact with electrodes, thus eliminating possible sources of contamination and allowing the operation of such plasma torches with a wide range of gases including inert, reducing, oxidising and other corrosive atmospheres. The working frequency is typically between 200 kHz and 40 MHz. Laboratory units run at power levels in the order of 30-50 kW while the large scale industrial units have been tested at power levels up to 1 MW.

As the residence time of the injected feed droplets in the plasma is very short it is important that the droplet sizes are small enough in order to obtain complete evaporation. The RF plasma method has been used to synthesize different nanoparticle materials, for example synthesis of various ceramic nanoparticles such as oxides, carbours/carbides and nitrides of Ti and Si.

Inert-gas aggregation is frequently used to make nanoparticles from metals with low melting points. The metal is vaporized in a vacuum chamber and then supercooled with an inert gas stream. The supercooled metal vapour condenses in to nanometer-sized particles, which can be entrained in the inert gas stream and deposited on a substrate or studied in situ.

Morphology

Scientists have taken to naming their particles after the real world shapes that they might represent. Nanospheres, nanoreefs, nanoboxes and more have appeared in the literature. These morphologies sometimes arise spontaneously as an effect of a templating or directing agent present in the synthesis such as micellular emulsions or anodized alumina pores, or from the innate crystallographic growth patterns of the materials themselves. Some of these morphologies may serve a purpose, such as long carbon nanotubes being used to bridge an electrical junction, or just a scientific curiosity like the stars shown at left.

Safety Issues

Nanoparticles present possible dangers, both medically and environmentally. Most of these are due to the high surface to volume ratio, which can make the particles very reactive or catalytic. They are also able to pass through cell membranes in organisms, and their interactions with biological systems are relatively unknown. However, free nanoparticles in the environment quickly tend to agglomerate and thus leave the nano-regime, and nature itself presents many nanoparticles to which organisms on earth may have evolved immunity.

According to the San Francisco Chronicle, "Animal studies have shown that some nanoparticles can penetrate cells and tissues, move through the body and brain and cause biochemical damage they also have shown to cause a risk factor in men for testicular cancer. But whether cosmetics and sunscreens containing nanomaterials pose health risks remains largely unknown, pending completion of long-range studies recently begun by the FDA and other agencies." Diesel nanoparticles have been found to damage the cardiovascular system in a mouse model.

Silicon Nanoparticle Cell

Generally, solar cells on the market today do not produce much electricity from ultraviolet light, instead it is either filtered out or absorbed by the cell, heating the cell. That heat is wasted energy and could even lead to damage to the cell. By diluting particles of silicon in alcohol, covering a solar cell with it and letting the alcohol evaporate to leave the nanoparticles of silicon on the cell has been increased the cell power output by 67% in the ultraviolet range and about 10% in the visible range.

Preparation of Nanoparticles

Nanoparticles can be prepared from a variety of materials such as proteins, polysaccharides and synthetic polymers. The selection of matrix materials is dependent on many factors including:

(a) size of nanoparticles required;

(b) inherent properties of the drug, e.g., aqueous solubility and stability;

(c) surface characteristics such as charge and permeability;

(d) degree of biodegradability, biocompatibility and toxicity;

(e) Drug release profile desired; and

(f) Antigenicity of the final product.

Nanoparticles have been prepared most frequency by three methods:

(1) dispersion of preformed polymers;

(2) polymerisation of monomers; and

(3) ionic gelation or coacervation of hydrophilic polymers.

However, other methods such as supercritical fluid technology and particle replication in non-wetting templates (PRINT) have also been described in the literature for production of nanoparticles. The latter was claimed to have absolute control of particle size, shape and composition, which could set an example for the future mass production of nanoparticles in industry.

— *Dispersion of preformed polymers*: Dispersion of preformed polymers is a common technique used to prepare biodegradable nanoparticles from poly (lactic acid) (PLA); poly (D,L-glycolide), PLG; poly (D, L-lactide-co-glycolide) (PLGA) and poly (cyanoacrylate) (PCA).

— *Solvent evaporation method*: In this method, the polymer is dissolved in an organic solvent such as dichloromethane, chloroform or ethyl acetate which is also used as the solvent for dissolving the hydrophobic drug. The mixture of polymer and drug solution is then emulsified in an aqueous solution containing a surfactant or emulsifying agent to form an oil in water (o/w) emulsion. After the formation of stable emulsion, the organic solvent is evaporated either by reducing the pressure or by continuous stirring. Particle size was found to be influenced by the type and concentrations of stabiliser, homogenizer speed and polymer concentration. In order to produce small particle size, often a high-speed homogenisation or ultrasonication may be employed.

— *Spontaneous emulsification or solvent diffusion method*: This is a modified version of solvent evaporation method. In this method, the watermiscible solvent along with a small amount of the water immiscible organic solvent is used as an oil phase. Due to the spontaneous diffusion of solvents an interfacial turbulence is created between the two phases leading to the formation of small particles. As the concentration of water miscible solvent increases, a decrease in the size of particle can be achieved.

Both solvent evaporation and solvent diffusion methods can be used for hydrophobic or hydrophilic drugs. In the case of hydrophilic drug, a multiple w/o/w emulsion needs to be formed with the drug dissolved in the internal aqueous phase.

Polymerisation Method

In this method, monomers are polymerised to form nanoparticles in an aqueous solution. Drug is incorporated either by being dissolved in the polymerisation medium or by adsorption onto the nanoparticles after polymerisation completed. The nanoparticle suspension is then purified to remove various stabilisers and surfactants employed for polymerisation by ultracentrifugation and re-suspending the particles in an isotonic surfactant-free medium. This technique has been reported for making polybutylcyanoacrylate or poly (alkylcyanoacrylate) nanoparticles;. Nanocapsule formation and their particle size depends on the concentration of the surfactants and stabilisers used.

Ionic Gelation Method

Much research has been focused on the preparation of nanoparticles using biodegradable hydrophilic polymers such as chitosan, gelatin and sodium alginate. Calvo and co-workers developed a method for preparing hydrophilic chitosan nanoparticles by ionic gelation. The method involves a mixture of two aqueous phases, of which one is the polymer chitosan, a di-block co-polymer ethylene oxide or propylene oxide (PEO-PPO) and the other is a polyanion sodium tripolyphosphate.

In this method, positively charged amino group of chitosan interacts with negative charged tripolyphosphate to form coacervates with a size in the range of nanometer. Coacervates are formed as a result of electrostatic interaction between two aqueous phases, whereas, ionic gelation involves the material undergoing transition from liquid to gel due to ionic interaction conditions at room temperature.

Supercritical Fluid Technology

Conventional methods such as solvent extraction-evaporation, solvent diffusion and organic phase separation methods require the use of organic solvents which are hazardous to the environment as well as to physiological systems. Therefore, the supercritical fluid technology has been investigated as an alternative to prepare biodegradable micro- and nanoparticles because supercritical fluids are environmentally safe.

A supercritical fluid can be generally defined as a solvent at a temperature above its critical temperature, at which the fluid remains a single phase regardless of pressure. Supercritical CO_2 (SC CO_2) is the most widely used supercritical fluid because of its mild critical conditions (T_c = 31.1 °C, P_c = 73.8 bars), nontoxicity, non-flammability, and low price. The most common processing techniques involving supercritical fluids are supercritical anti-solvent (SAS) and rapid expansion of critical solution (RESS).

The process of SAS employs a liquid solvent, eg methanol, which is completely miscible with the supercritical fluid (SC CO_2), to dissolve the solute to be micronized; at the process conditions, because the solute is insoluble in the supercritical fluid, the extract of the liquid solvent by supercritical fluid leads to the instantaneous precipitation of the solute, resulting the formation of nanoparticles. Thote and Gupta reported the use of a modified SAS method for formation of hydrophilic drug dexamethasone phosphate drug nanoparticles for microencapsulation purpose.

RESS differs from the SAS process in that its solute is dissolved in a supercritical fluid (such as supercritical methanol) and then the solution is rapidly expanded through a small nozzle into a region lower pressure, Thus the solvent power of supercritical fluids dramatically decreases and the solute eventually precipitates. This technique is clean because the precipitate is basically solvent free. RESS and its modified process have been used for the product of polymeric nanoparticles. Supercritical fluid technology technique, although environmentally friendly and suitable for mass production, requires specially designed equipment and is more expensive.

NANOPARTICLES ON DRUG DELIVERY

Particle Size

Particle size and size distribution are the most important characteristics of nanoparticle systems. They determine the in vivo distribution, biological fate, toxicity and the targeting ability of nanoparticle systems. In addition, they can also influence the drug loading, drug release and stability of nanoparticles. Many studies have demonstrated that nanoparticles of sub-micron size have a number of advantages over microparticles as a drug delivery system.

Generally nanoparticles have relatively higher intracellular uptake compared to microparticles and available to a wider range of biological targets due to their small size and relative mobility. Desai et al found that 100 nm nanoparticles had a 2.5 fold greater uptake than 1 μm microparticles, and 6 fold greater uptake than 10 μm microparticles in a Caco-2 cell line.

In a subsequent study, the nanoparticles penetrated throughout the submucosal layers in a rat in situ intestinal loop model, while microparticles were predominantly localised in the epithelial lining. It was also reported that nanoparticles can across the blood-brain barrier following the opening of tight junctions by hyper osmotic mannitol, which may provide sustained delivery of therapeutic agents for difficult-to-treat diseases like brain tumours. Tween 80 coated nanoparticles have been shown to cross the blood-brain barrier. In some cell lines, only submicron nanoparticles can be taken up efficiently but not the larger size microparticles.

Drug release is affected by particle size. Smaller particles have larger surface area, therefore, most of the drug associated would be at or near the particle surface, leading to fast drug release. Whereas, larger particles have large cores which allow more drug to be encapsulated and slowly diffuse out. Smaller particles also have greater risk of aggregation of particles during storage and transportation of nanoparticle dispersion. It is always a challenge to formulate nanoparticles with the smallest size possible but maximum stability.

Polymer degradation can also be affected by the particle size. For instance, the rate of PLGA polymer degradation was found to increase with increasing particle size in vitro. It was thought that in smaller particles, degradation products of PLGA formed can diffuse out of the particles easily while in large particles, degradation products are more likely remained within the polymer matrix for a longer period to cause autocatalytic degradation of the polymer material. Therefore, it was hypothesized that larger particles will contribute to faster polymer degradation as well as the drug release.

However, Panyam et al prepared PLGA particles with different size ranges and found that the polymer degradation rates in vitro were not substantially different for different size particles. Currently, the fastest and most routine method of determining particle size is by photon-correlation spectroscopy or dynamic light scattering. Photon-correlation spectroscopy requires the viscosity of the medium to be known and determines the diameter of the particle by Brownian motion and light scattering properties. The results obtained by photon-correlation spectroscopy are usually verified by scanning or transmission electron microscopy (SEM or TEM).

Surface Properties of Nanoparticles

When nanoparticles are administered intravenously, they are easily recognised by the body immune systems, and are then cleared by phagocytes from the circulation. Apart from the size of nanoparticles, their surface hydrophobicity determines the amount of adsorbed blood components, mainly proteins (opsonins). This in turn influences the in vivo fate of nanoparticles. Binding of these opsonins onto the surface of nanoparticles called opsonisation acts as a bridge between nanoparticles and phagocytes.

The association of a drug to conventional carriers leads to modification of the drug biodistribution profile, as it is mainly delivered to the mononuclear phagocytes system (MPS) such as liver, spleen, lungs and bone marrow. Indeed, once in the blood stream, surface non-modified nanoparticles (conventional nanoparticles) are rapidly opsonised and massively cleared by the macrophages of MPS rich organs.

Generally, it is IgG, compliment C_3 components that are used for recognition of foreign substances, especially foreign macromolecules.

Hence, to increase the likelihood of the success in drug targeting by nanoparticles, it is necessary to minimize the opsonisation and to prolong the circulation of nanoparticles in vivo. This can be achieved by

(a) surface coating of nanoparticles with hydrophilic polymers/surfactants;

(b) formulation of nanoparticles with biodegradable copolymers with hydrophilic segments such as polyethylene glycol (PEG), polyethylene oxide, polyoxamer, poloxamine and polysorbate 80.

Studies show that PEG conformation at the nanoparticle surface is of utmost importance for the opsonin repelling function of the PEG layer. PEG surfaces in brush-like and intermediate configurations reduced phagocytosis and complement activation whereas PEG surfaces in mushroom-like configuration were potent complement activators and favoured phagocytosis.

The zeta potential of a nanoparticle is commonly used to characterise the surface charge property of nanoparticles. It reflects the electrical potential of particles and is influenced by the composition of the particle and the medium in which it is dispersed. Nanoparticles with a zeta potential above (+/-) 30 mV have been shown to be stable in suspension, as the surface charge prevents aggregation of the particles. The zeta potential can also be used to determine whether a charged active material is encapsulated within the centre of the nanocapsule or adsorbed onto the surface.

Drug Loading Capacity

Ideally, a successful nanoparticulate system should have a high drug-loading capacity thereby reduce the quantity of matrix materials for administration. Drug loading can be done by two methods:

— Incorporating at the time of nanoparticles production (incorporation method)

— Absorbing the drug after formation of nanoparticles by incubating the carrier with a concentrated drug solution

Drug loading and entrapment efficiency very much depend on the solid-state drug solubility in matrix material or polymer, which is related to the polymer composition, the molecular weight, the drug polymer interaction and the presence of endfunctional groups. The PEG moiety has no or little effect on drug loading. The macromolecule or protein shows greatest loading efficiency when it is loaded at or near its isoelectric point when it has minimum solubility and maximum adsorption For small molecules, studies show the use of ionic interaction between the drug and matrix materials can be a very effective way to increase the drug loading.

Drug release

To develop a successful nanoparticulate system, both drug release and polymer biodegradation are important consideration factors. In general, drug release rate depends on:

(1) solubility of drug;

(2) desorption of the surfacebound/ adsorbed drug;

(3) drug diffusion through the nanoparticle matrix;

(4) nanoparticle matrix erosion/degradation; and

(5) combination of erosion/diffusion process.

Thus solubility, diffusion and biodegradation of the matrix materials govern the release process. In the case of nanospheres, where the drug is uniformly distributed, the release occurs by diffusion or erosion of the matrix under sink conditions. If the diffusion of the drug is faster than matrix erosion, the mechanism of release is largely controlled by a diffusion process.

The rapid initial release or 'burst' is mainly attributed to weakly bound or adsorbed drug to the large surface of nanoparticles. It is evident that the method of incorporation has an effect on release profile. If the drug is loaded by incorporation method, the system has a relatively small burst effect and better sustained release characteristics. If the nanoparticle is coated by polymer, the release is then controlled by diffusion of the drug from the core across the polymeric membrane. The membrane coating acts as a barrier to release, therefore, the solubility and diffusivity of drug in polymer membrane becomes determining factor in drug release. Furthermore release rate can also be affected by ionic interaction between the drug and addition of auxillary ingredients. When the drug is involved in interaction with auxillary ingredients to form a less water soluble complex, then the drug release can be very slow with almost no burst release effect ; whereas if the addition of auxillary ingredients e.g., addition of ethylene oxide-propylene oxide block copolymer (PEO-PPO) to chitosan, reduces the interaction of the model drug bovine serum albumin (BSA) with the matrix material (chitosan) due to competitive electrostatic interaction of PEO-PPO with chitosan, then an increase in drug release could be observed.

Various methods which can be used to study the in vitro release of the drug are:

(1) side-by-side diffusion cells with artificial or biological membranes;

(2) dialysis bag diffusion technique;

(3) reverse dialysis bag technique;

(4) agitation followed by ultracentrifugation/centrifugation;

(5) Ultra-filtration or centrifugal ultra-filtration techniques.

Usually the release study is carried out by controlled agitation followed by centrifugation. Due to the time-consuming nature and technical difficulties encountered in the separation of nanoparticles from release media, the dialysis technique is generally preferred.

NANOPARTICULATE DELIVERY SYSTEMS

Tumour Targeting

The rationale of using nanoparticles for tumour targeting is based on

1) nanoparticles will be able to deliver a concentrate dose of drug in the vicinity of the tumour targets via the enhanced permeability and retention effect or active targeting by ligands on the surface of nanoparticles;

2) nanoparticles will reduce the drug exposure of health tissues by limiting drug distribution to target organ.

Verdun et al demonstrated in mice treated with doxorubicin incorporated into poly nanopsheres that higher concentrations of doxorubicin manifested in the liver, spleen and lungs than in mice treated with free doxorubicin. Studies show that the polymeric composition of nanoparticles such as type, hydrophobicity and biodegradation profile of the polymer along with the associated drug's molecular weight, its localisation in the nanospheres and mode of incorporation technique, adsorption or incorporation, have a great influence on the drug distribution pattern in vivo.

The exact underlying mechanism is not fully understood but the biodistribution of nanoparticles is rapid, within ½ hour to 3 hours, and it likely involves MPS and endocytosis/phagocytosis process. Recently Bibby et al reported the biodistribution and pharmacokinetics (PK) of a cyclic RGD doxorubicin-nanoparticle formulation in tumour bearing mice. Their biodistribution studies revealed decreasing drug concentrations over time in the heart, lung, kidney and plasma and accumulating drug concentrations in the liver, spleen and tumour.

The majority injected dose appeared in the liver (56%) and only 1.6% in the tumour at 48 hrs post injection, confirming that nanoparticles have a great tendency to be captured by liver. This indicates the greatest challenge of using nanoparticles for tumour targeting is to avoid particle uptake by mononuclear phagocytic system (MPS) in liver and spleen.

Such propensity of MPS for endocytosis/phagocytosis of nanoparticles provides an opportunity to effectively deliver therapeutic agents to these cells. This biodistribution can be of benefit for the chemotherapeutic treatment of MPS- rich organs/tissues localised tumours like hepatocarcinoma, hepatic metastasis arising from digestive tract or gynaecological cancers, brochopulmonary tumours, primitive tumours and metastasis, small cell tumours, myeloma and leukaemia.

It has been proved that using doxorubicin loaded conventional nanoparticles was effective against hepatic metastasis model in mice. It was found there was greater reduction in the degree of metastasis than when free drug was used. The underlying mechanism responsible for the increased therapeutic efficacy of the formulation was transfer of doxorubicin from healthy tissue, acting as a drug reservoir to the malignant tissues. Histological examination showed a considerable accumulation of nanoparticles in the lysosomal vesicles of Kupffer cells, whereas nanoparticles could not be clearly identified in tumoral cells. Thus Kupffer cells, after a massive uptake of nanoparticles by phagocytosis, were able to induce the release of doxorubicin, leading to a gradient of drug concentration, favourable for a prolonged diffusion of the free and still active drug towards the neighbouring metastatic cells.

When conventional nanoparticles are used as carriers in chemotherapy, some cytotoxicity against the Kupffer cells can be expected, which would result in deficiency of Kupffer cells and naturally lead to reduced liver uptake and decreased therapeutic effect with intervals of less than 2 weeks administration.

Moreover, conventional nanoparticles can also target bone marrow (MPS tissue), which is an important but unfavourable site of action for most anticancer drugs because chemotherapy with such carriers may increase myelosuppresive effect. Therefore, the ability of conventional nanoparticles to enhance anticancer drugs efficacy is limited to targeting tumours at the level of MPS-rich organs. Also, directing anticancer drug-loaded nanoparticles to other tumoral sites is not feasible if a rapid clearance of nanoparticles occurs shortly after intravenous administration.

Long Circulating Nanoparticles

To be successful as a drug delivery system, nanoparticles must be able to target tumours which are localised outside MPS-rich organs. In the past decade, a great deal of work has been devoted to developing so-called "stealth" particles or PEGylated nanoparticles, which are invisible to macrophages or phagocytes. A major breakthrough in the field came when the use of hydrophilic polymers to efficiently coat conventional nanoparticle surface produced an opposing effect to the uptake by the MPS.

These coatings provide a dynamic "cloud" of hydrophilic and neutral chains at the particle surface which repel plasma proteins. As a result, those coated nanoparticles become invisible to MPS, therefore, remained in the circulation for a longer period of time. Hydrophilic polymers can be introduced at the surface in two ways, either by adsorption of surfactants or by use of block or branched copolymers for production of nanoparticles.

Studies show nanoparticles containing a coat of PEG not only have a prolonged half-life in the blood compartment but also be able to selectively extravasate in pathological sites such as tumours or inflamed regions with a leaky vasculature. As a result, such long-circulating nanoparticles have increased the potential to directly target tumours located outside MPS-rich regions. The size of the colloidal carriers as well as their surface characteristics are the critical to the biological fate of nanoparticles. A size less than 100 nm and a hydrophilic surface are essential in achieving the reduction of opsonisation reactions and subsequent clearance by macrophages.

Coating conventional nanoparticles with surfactants or PEG to obtain a long-circulating carrier has now been used as a standard strategy for drug targeting in vivo. Extensive efforts have been devoted to achieving "active targeting" of nanoparticles in order to deliver drugs to the right targets, based on molecular recognition processes such as ligand-receptor or antigen-antibody interaction.

Considering that fact that folate receptors are over expressed on the surface of some human malignant cells and the cell adhesion molecules such as selectins and integrins are involved in metastatic events, nanoparticles bearing specific ligands such as folate may be used to target ovarian carcinoma while specific peptides or carbohydrates may be used to target integrins and selectins. Oyewumi et al demonstrated that the benefits of folate ligand coating were to facilitate tumour cell internalisation and retention of Gd-nanoparticles in the tumour tissue.

Targeting with small ligands appears more likely to succeed since they are easier to handle and manufacture. Furthermore, it could be advantageous when the active targeting ligands are used in combination with the long-circulating nanoparticles to maximize the likelihood of the success in active targeting of nanoparticles.

Multidrug Resistance in Tumour Cells

Anticancer drugs, even if they are located in the tumour interstitium, can turn out to be of limited efficacy against numerous solid tumour types, because cancer cells are able to develop mechanisms of resistance. These mechanisms allow tumours to evade chemotherapy. Multidrug resistance (MDR) is one of the most serious problems in chemotherapy. MDR occurs mainly due to the over expression of the plasma membrane pglycoprotein (Pg_p), which is capable of extruding various positively charged xenobiotics, including some anticancer drugs, out of cells.

In order to restore the tumoural cells' sensitivity to anticancer drugs by circumventing Pg_p-mediated MDR, several strategies including the use of colloidal carriers have been applied. The rationale behind the association of drugs with colloidal carriers, such as nanoparticles, against drug resistance derives from the fact that Pgprobably recognises the drug to be effluxed out of the tumoural cells only when this drug is present in the plasma membrane, and not when it is located in the cytoplasm or lysosomes after endocytosis.

Oral Delivery of Peptides and Proteins

Significant advances in biotechnology and biochemistry have led to the discovery of a large number of bioactive molecules and vaccines based on peptides and proteins. Development of suitable carriers remains a challenge due to the fact that bioavailability of these molecules is limited by the epithelial barriers of the gastrointestinal tract and their susceptibility to gastrointestinal degradation by digestive enzymes. Polymeric nanoparticles allow encapsulation of bioactive molecules and protect them against enzymatic and hydrolytic degradation. For instance, it has been found that insulin-loaded nanoparticles have preserved insulin activity and produced blood glucose reduction in diabetic rats for up to 14 days following the oral administration.

The surface area of human mucosa extends to 200 times that of skin. The gastrointestinal tract provides a variety of physiological and morphological barriers against protein or peptide delivery, e.g.,

(a) proteolytic enzymes in the gut lumen like pepsin, trypsin and chymotrypsin;

(b) proteolytic enzymes at the brush border membrane (endopeptidases);

(c) bacterial gut flora; and

(d) mucus layer and epithelial cell lining itself.

The histological architecture of the mucosa is designed to efficiently prevent uptake of particulate matter from the environment. One important strategy to overcome the gastrointestinal barrier is to deliver the drug in a colloidal carrier system, such as nanoparticles, which is capable of enhancing the interaction mechanisms of the drug delivery system and the epithelia cells in the GI tract.

Epithelial Cells in the GI Tract Using Ligands

Targeting strategies to improve the interaction of nanoparticles with adsorptive enterocytes and M-cells of Peyer's patches in the GI tract can be classified into those utilising specific binding to ligands or receptors and those based on nonspecific adsorptive mechanism. The surface of enterocytes and M cells display cell-specific carbohydrates, which may serve as binding sites to colloidal drug carriers containing appropriate ligands. Certain glycoproteins and lectins bind selectively to this type of surface structure by specific receptor-mediated mechanism. Different lectins, such as bean lectin and tomato lectin, have been studied to enhance oral peptide adsorption.

Vitamin B-12 absorption from the gut under physiological conditions occurs via receptor-mediated endocytosis. The ability to increase oral bioavailability of various peptides and particles by covalent coupling to vitamin B-12 has been studied. For this intrinsic process, mucoprotein is required, which is prepared by the mucus membrane in the stomach and binds specifically to cobalamin. The mucoprotein completely reaches the ileum where resorption is mediated by specific receptors.

Absorption of nanoparticles

In general, the gastrointestinal absorption of macromolecules and particulate materials involves either paracellular route or endocytotic pathway. The paracellular route of absorption of nanoparticles utilises less than 1% of mucosal surface area. Using polymers such as chitosan, starch or poly (acrylate) can increase the paracellular permeability of macromolecules. Endocytotic pathway for absorption of nanoparticles is either by receptor-mediated endocytosis, that is, active targeting, or adsorptive endocytosis which does not need any ligands. This process is initiated by an unspecific physical adsorption of material to the cell surface by electrostatic forces such as hydrogen bonding or hydrophobic interactions.

Adsorptive endocytosis depends primarily on the size and surface properties of the material. If the surface charge of the nanoparticles is positive or uncharged, it will provide an affinity to adsorptive enterocytes though hydrophobic, whereas if it is negatively charged and hydrophilic, it shows greater affinity to adsorptive enterocytes and Mcells. This shows that a combination of size, surface charge and hydrophilicity play a major role in affinity. This is demonstrated with poly (styrene) nanoparticles and when it is carboxylated.

Nanoparticles for Gene Delivery

Polynucleotide vaccines work by delivering genes encoding relevant antigens to host cells where they are expressed, producing the antigenic protein within the vicinity of professional antigen presenting cells to initiate immune response. Such vaccines produce both humoral and cell-mediated immunity because intracellular production of protein, as opposed to extracellular deposition, stimulates both arms of the immune system.

The key ingredient of polynucleotide vaccines, DNA, can be produced cheaply and has much better storage and handling properties than the ingredients of the majority of protein-based vaccines. Hence, polynucleotide vaccines are set to supersede many conventional vaccines particularly for immunotherapy. However, there are several issues related to the delivery of polynucleotides which limit their application. These issues include efficient delivery of the polynucleotide to the target cell population and its localisation to the nucleus of these cells, and ensuring that the integrity of the polynucleotide is maintained during delivery to the target site.

Nanoparticles loaded with plasmid DNA could also serve as an efficient sustained release gene delivery system due to their rapid escape from the degradative endo-lysosomal compartment to the cytoplasmic compartment. Hedley et al. reported that following their intracellular uptake and endolysosomal escape, nanoparticles could release DNA at a sustained rate resulting in sustained gene expression. This gene delivery strategy could be applied to facilitate bone healing by using PLGA nanoparticles containing therapeutic genes such as bone morphogenic protein.

Blood-Brain Barrier

The blood-brain barrier (BBB) is the most important factor limiting the development of new drugs for the central nervous system. The BBB is characterised by relatively impermeable endothelial cells with tight junctions, enzymatic activity and active efflux transport systems. It effectively prevents the passage of water-soluble molecules from the blood circulation into the CNS, and can also reduce the brain concentration of lipid-soluble molecules by the function of enzymes or efflux pumps.

Consequently, the BBB only permits selective transport of molecules that are essential for brain function. Strategies for nanoparticle targeting to the brain rely on the presence of and nanoparticle interaction with specific receptor-mediated transport systems in the BBB. For example polysorbate 80/LDL, transferrin receptor binding antibody (such as OX26), lactoferrin, cellpenetrating peptides and melanotransferrin have been shown capable of delivery of a self non transportable drug into the brain via the chimeric construct that can undergo receptor-mediated transcytosis.

It has been reported poly nanoparticles was able to deliver hexapeptide dalargin, doxorubicin and other agents into the brain which is significant because of the great difficulty for drugs to cross the BBB. Despite some reported success with polysorbate 80 coated NPs, this system does have many shortcomings including desorption of polysorbate coating, rapid NP degradation and toxicity caused by presence of high concentration of polysorbate 80.. OX26 MAbs (anti-transferrin receptor MAbs), the most studied BBB targeting antibody, have been used to enhance the BBB penetration of lipsosomes.

References

Buffat, Ph.; Burrel, J.-P., "Size effect on the melting temperature of gold particles", *Physical Review* A 13 (6): 2287-2298, 1976.

Choy JH, Jang ES, Won JH, Chung JH, Jang DJ, and Kim YW., "Hydrothermal route to ZnO nanocoral reefs and nanofibers", *Appl. Phys. Lett.* 84 287, 2004.

6

Carbon Nanotubes

Many exotic structures of fullerenes exist: regular spheres, cones, tubes and also more complicated and strange shapes. Single Walled Nanotubes (SWNT) can be considered as long wrapped graphene sheets. Nanotubes generally have a length to diameter ratio of about 1000 so they can be considered as nearly one-dimensional structures. A SWNT consists of two separate regions with different physical and chemical properties.

The first is the sidewall of the tube and the second is the end cap of the tube. The end cap structure is similar to or derived from a smaller fullerene, such as C_{60}. C-atoms placed in hexagons and pentagons form the end cap structures. It can be easily derived from Euler's theorem that twelve pentagons are needed in order to obtain a closed cage structure which consists of only pentagons and hexagons. The combination of a pentagon and five surrounding hexagons results in the desired curvature of the surface to enclose a volume.

A second rule is the isolated pentagon rule that states that the distance between pentagons on the fullerene shell is maximised in order to obtain a minimal local curvature and surface stress, resulting in a more stable structure. The smallest stable structure that can be made this way is C_{60} the one just larger is C_{70} and so on.

Another property is that all fullerenes are composed of an even number of C-atoms because adding one hexagon to an existing structure means adding two C-atoms. The other structure of which a SWNT is composed is a cylinder. It is generated when a graphene sheet of a certain size that is wrapped in a certain direction. As the result is cylinder symmetric we can only roll in a discreet set of directions in order to form a closed cylinder.

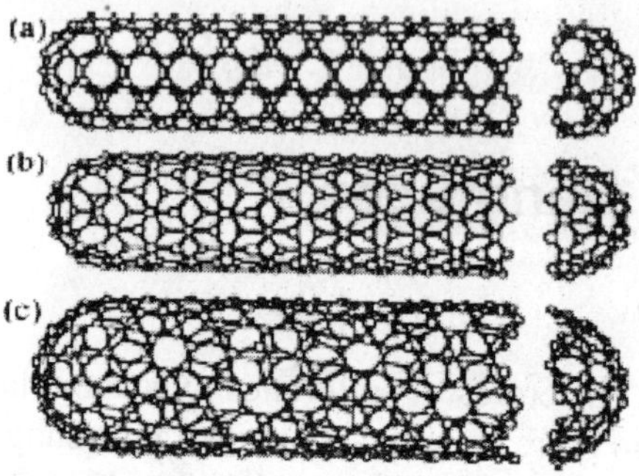

Figure 1: Some SWNTs with different chiralities.

Two atoms in the graphene sheet are chosen, one of which servers the role as origin. The sheet is rolled until the two atoms coincide. The vector pointing from the first atom towards the other is called the chiral vector and its length is equal to the circumference of the nanotube. (Figure 1) The direction of the nanotube axis is perpendicular to the chiral vector. SWNTs with different chiral vectors have dissimilar properties such as optical activity, mechanical strength and electrical conductivity.

After ideal structures without flaws. Deformations, such as bends and nanotube junctions, are introduced by replacing a hexagon with a heptagon or pentagon. Deformations can be inward or outward and, among others, electrical properties are seriously changed by these deformations. Another class of defects is caused by impurities that are built in during or after the nanotube growth process; Compounds that can be incorporated into the structure are for example catalyst particles.

Multi Walled Nanotubes (MWNT) can be considered as a collection of concentric SWNTs with different diameters. The length and diameter of these structures differ a lot from those of SWNTs and, of course, their properties are also very different. It can be considered as a gradual transition from a large diameter to a smaller one without defects in the wall of the cone but with fewer pentagons in the end cap.

Introduction of defects can also result in various new structures such as Y-branches T-branches or SWNT junctions. Under certain circumstances, these defects can be introduced in a 'controlled' way. These defects result in special structures that will have other, but even more interesting, properties than their original forms. Defects are also being studied with great effort. A final type of interesting structures is the so-called peapods, carbon nanotubes with C_{60} molecules enclosed in the nanotube.

Structural Properties of Carbon Nanotubes

Electronic, molecular and structural properties of carbon nanotubes are determined to a large extent by their nearly one dimensional structure. The most important properties of CNTs and their molecular background are stated below.

— *Chemical reactivity:* The chemical reactivity of a CNT is, compared with a graphene sheet, enhanced as a direct result of the curvature of the CNT surface. Carbon nanotube reactivity is directly related to the pi-orbital mismatch caused by an increased curvature. Therefore, a distinction must be made between the sidewall and the end caps of a nanotube. For the same reason, a smaller nanotube diameter results in increased reactivity. Covalent chemical modification of either sidewalls or end caps has shown to be possible. For example, the solubility of CNTs in different solvents can be controlled this way. Though, direct investigation of chemical modifications on nanotube behaviour is difficult as the crude nanotube samples are still not pure enough.

— *Electrical conductivity:* Depending on their chiral vector, carbon nanotubes with a small diameter are either semi-conducting or metallic. The differences in conducting properties are caused by the molecular structure that results in a different band structure and thus a different band gap. The differences in conductivity can easily be derived from the graphene sheet properties. It was shown that a (n,m) nanotube is metallic as accounts that: n = m or (n-m) = 3i, where i is an integer and n and m are defining the nanotube. The resistance to conduction is determined by quantum mechanical aspects and was proved to be independent of the nanotube length.

— *Optical activity:* Theoretical studies have revealed that the optical activity of chiral nanotubes disappears if the nanotubes become larger. Therefore, it is expected that other physical properties are influenced by these parameters too. Use of the optical activity might result in optical devices in which CNTs play an important role.

— *Mechanical strength:* Carbon nanotubes have a very large Young modulus in their axial direction. The nanotube as a whole is very flexible because of the great length. Therefore, these compounds are potentially suitable for applications in composite materials that need anisotropic properties.

Techniques for Nanotube Synthesis

First, the growth mechanism is explained, as it is almost general for all techniques. However, typical conditions are stated at the sections of all the different techniques. The largest interest is in the newest methods for each technique and the possibilities of scaling up. Carbon nanotubes are generally produced by three main techniques, arc discharge, laser ablation and chemical vapour deposition. Though scientists are researching more economic ways to produce these structures. In arc discharge, a vapour is created by an arc discharge between two carbon electrodes with or without catalyst.

Nanotubes self-assemble from the resulting carbon vapour. In the laser ablation technique, a high-power laser beam impinges on a volume of carbon -containing feedstock gas (methane or carbon monoxide). At the moment, laser ablation produces a small amount of clean nanotubes, whereas arc discharge methods generally produce large quantities of impure material. In general, chemical vapour deposition (CVD) results in MWNTs or poor quality SWNTs. The SWNTs produced with CVD have a large diameter range, which can be poorly controlled. But on the other hand, this method is very easy to scale up, what favours commercial production.

Growth mechanism

The way in which nanotubes are formed is not exactly known. The growth mechanism is still a subject of controversy, and more than one mechanism might be operative during the formation of CNTs. One of the mechanisms consists out of three steps.

First a precursor to the formation of nanotubes and fullerenes, C_2, is formed on the surface of the metal catalyst particle. From this metastable carbide particle, a rodlike carbon is formed rapidly. Secondly there is a slow graphitisation of its wall. This mechanism is based on in-situ TEM observations. The exact atmospheric conditions depend on the technique used, later on, these will be explained for each technique as they are specific for a technique. The actual growth of the nanotube seems to be the same for all techniques mentioned.

There are several theories on the exact growth mechanism for nanotubes. One theory postulates that metal catalyst particles are floating or are supported on graphite or another substrate. It presumes that the catalyst particles are spherical or pear-shaped, in which case the deposition will take place on only one half of the surface. The carbon diffuses along the concentration gradient and precipitates on the opposite half, around and below the bisecting diameter. However, it does not precipitate from the apex of the hemisphere, which accounts for the hollow core that is characteristic of these filaments.

For supported metals, filaments can form either by 'extrusion' in which the nanotube grows upwards from the metal particles that remain attached to the substrate, or the particles detach and move at the head of the growing nanotube, labelled 'tip-growth'. Depending on the size of the catalyst particles, SWNT or MWNT are grown. In arc discharge, if no catalyst is present in the graphite, MWNT will be grown on the C_2-particles that are formed in the plasma.

Arc Discharge Method

The carbon arc discharge method, initially used for producing C_{60} fullerenes, is the most common and perhaps easiest way to produce carbon nanotubes as it is rather simple to undertake. However, it is a technique that produces a mixture of components and requires separating nanotubes from the soot and the catalytic metals present in the crude product. This method creates nanotubes through arc-vaporisation of two carbon rods placed end to end, separated by approximately 1 mm, in an enclosure that is usually filled with inert gas (helium, argon) at low pressure (between 50 and 700 mbar).

Recent investigations have shown that it is also possible to create nanotubes with the arc method in liquid nitrogen. A direct current of 50 to 100 A driven by approximately 20 V creates a high temperature discharge between the two electrodes. The discharge vaporises one of the carbon rods and forms a small rod shaped deposit on the other rod. Producing nanotubes in high yield depends on the uniformity of the plasma arc and the temperature of the deposit form on the carbon electrode.

Insight in the growth mechanism is increasing and measurements have shown that different diameter distributions have been found depending on the mixture of helium and argon. These mixtures have different diffusions coefficients and thermal conductivities. These properties affect the speed with which the carbon and catalyst molecules diffuse and cool., affecting nanotube diameter in the arc process.

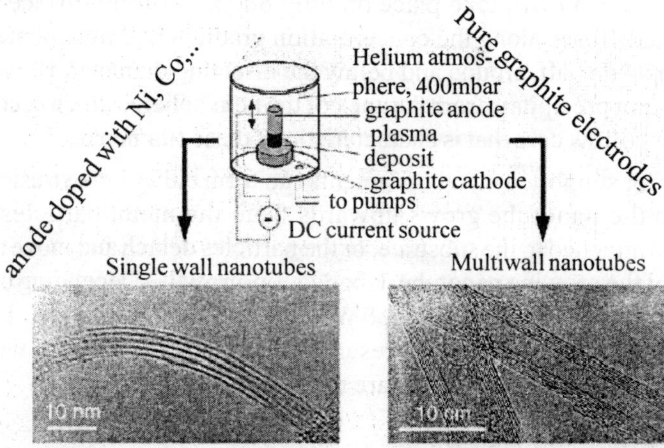

Figure 2: Experimental set-up of an arc discharge apparatus.

This implies that single-layer tubules nucleate and grow on metal particles in different sizes depending on the quenching rate in the plasma and it suggests that temperature and carbon and metal catalyst densities affect the diameter distribution of nanotubes. Depending on the exact technique, it is possible to selectively grow SWNTs or

MWNTs. Two distinct methods of synthesis can be performed with the arc discharge apparatus.

Synthesis of SWNT

If SWNTs are preferable, the anode has to be doped with metal catalyst, such as Fe, Co, Ni, Y or Mo. A lot of elements and mixtures of elements have been tested by various authors and it is noted that the results vary a lot, even though they use the same elements. This is not surprising as experimental conditions differ. The quantity and quality of the nanotubes obtained depend on various parameters such as the metal concentration, inert gas pressure, kind of gas, the current and system geometry. Usually the diameter is in the range of 1.2 to 1.4 nm. A couple of ways to improve the process of arc discharge are stated below.

- *Inert gas:* The most common problems with SWNT synthesis are that the product contains a lot of metal catalyst, SWNTs have defects and purification is hard to perform. On the other hand, an advantage is that the diameter can slightly be controlled by changing thermal transfer and diffusion, and hence condensation of atomic carbon and metals between the plasma and the vicinity of the cathode can control nanotube diameter in the arc process. This was shown in an experiment in which different mixtures of inert gases were used. It appeared that argon, with a lower thermal conductivity and diffusion coefficient, gave SWNTs with a smaller diameter of approximately 1.2 nm. A linear fit of the average nanotube diameter showed a 0.2 nm diameter decrease per 10 % increase in argon helium ratio, when nickel/yttrium was used (C/Ni/Y was 94.8:4.2:1) as catalyst.
- *Optical plasma control:* A second way of control is plasma control by changing the anode to cathode distance (ACD). The ACD is adjusted in order to obtain strong visible vortices around the cathode. This enhances anode vaporisation, which improves nanotubes formation. Combined with controlling the argon-helium mixture, one can simultaneously control the macroscopic and microscopic parameters of the nanotubes formed. With a nickel and yttrium catalyst (C/Ni/Y is 94.8:4.2:1) the optimum nanotube yield was found at a pressure of 660 mbar for pure

helium and 100 mbar for pure argon. The nanotube diameter ranges from 1.27 to 1.37 nanometre.

— *Catalyst:* Knowing that chemical vapour deposition (CVD) could give SWNTs with a diameter of 0.6-1.2 nm, researchers tried the same catalyst as used in CVD on arc discharge. Not all of the catalysts used appeared to result in nanotubes for both methods. But there seemed to be a correlation of diameter of SWNTs synthesised by CVD and arc discharge As a result, the diameter can be controllably lowered to a range of 0.6-1.2 nm with arc-discharge. Using a mixture of Co and Mo in high concentrations as catalyst resulted in this result. These diameters are considerably smaller than 1.2-1.4 nm, which is the usual size gained from arcdischarge.

— *Improvement of oxidation resistance:* There is also progress in developing methods to improve the oxidation resistance of the SWNTs, which is a consequence of the defects present in nanotubes. A strong oxidation resistance is needed for the nanotubes if they have to be used for applications such as field emission displays.

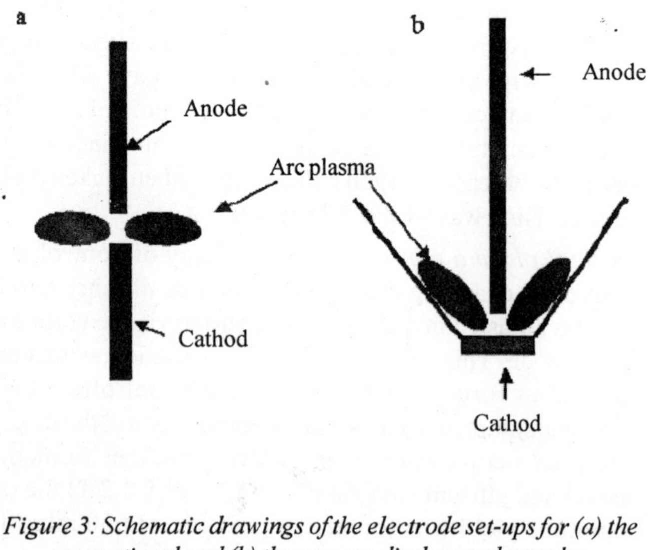

Figure 3: Schematic drawings of the electrode set-ups for (a) the conventional and (b) the new arc discharge electrodes.

Recent research has indicated that a modified arc-discharge method using a bowl-like cathode (Figure 3), decreases the defects and gives cleaner nanotubes, and thus improves the oxidation resistance. The Raman spectrum of the newly synthesised nanotubes shows that the nanotubes formed are cleaner and less defective compared with those synthesised by conventional methods. The anode rod contained Ni and Y catalyst (C /Ni/Y is 94.8:4.2:1). No information is given about the diameter size.

— *Open air synthesis with welding arc torch:* Only a couple of years ago, researchers discovered that it was possible to form MWNTs in open air. A welding arc torch was operated in open air and the process was shielded with an argon gas flow. The anode and cathode were made of graphite containing Ni and Y (Ni/Y is 4.2:1 at. %). This method was modified for preparing SWNTs. A plate target made of graphite containing metal catalyst Ni and Y (Ni/Y is 3.6:0.8 at. per cent), was fixed at the sidewall of a water-cooled, steel based electrode. The torch arc aimed at the edge of the target and the soot was deposited on the substrate behind the target. The arc was operated at a direct current of 100 A. and shielding argon gas flowed through the torch, enhancing the arc jet formation beyond the target.

In the soot, carbon nanohorns (CNHs) and bundles of SWNT with an average diameter of 1.32 nm were found. However, the yield was much lower than for the conventional low-pressure arc discharge method. There are two reasons for this fact. At first, because of the open air, the lighter soot will escape into the atmosphere. Secondly, the carbon vapour might be oxidised and emitted as carbon dioxide gas. In order to improve the yield in this method, contrivances for collecting soot and development of an appropriate target are required.

This method promises to be convenient and inexpensive once the conditions for higher yield are optimised. With a Ni/Y catalyst (Ni/Y is 3.6:0.8), SWNT bundles and CNHs are formed. In this case the SWNTs have a diameter of approximately 1.32 nm.

Synthesis of MWNT

If both electrodes are graphite, the main product will be MWNTs. But next to MWNTs a lot of side products are formed such as fullerenes, amorphous carbon, and some graphite sheets. Purifying the MWNTs, means loss of structure and disorders the walls. However scientist are developing ways to gain pure MWNTs in a large-scale process without purification. Typical sizes for MWNTs are an inner diameter of 1-3 nm and an outer diameter of approximately 10 nm. Because no catalyst is involved in this process, there is no need for a heavy acidic purification step. This means, the MWNT, can be synthesised with a low amount of defects.

Synthesis in liquid nitrogen

A first, possibly economical route to highly crystalline MWNTs is the arc-discharge method in liquid nitrogen, with this route mass production is also possible. For this option low pressures and expensive inert gasses are not needed. The content of the MWNTs can be as high as 70 % of the reaction product. Analysis with Augerspectroscopy revealed that no nitrogen was incorporated in the MWNTs. There is a strong possibility that SWNTs can be produced with the same apparatus under different conditions.

Magnetic field synthesis

Synthesis of MWNTs in a magnetic fieldgives defect-free and high purity MWNTs that can be applied as nanosized electric wires for device fabrication. In this case, the arc discharge synthesis was controlled by a magnetic field around the arc plasma. Extremely pure graphite rods (purity > 99.999 %) were used as electrodes. Highly pure MWNTs (purity > 95 %) were obtained without further purification, which disorders walls of MWNTs.

Plasma rotating arc discharge

A second possibly economical route to mass production of MWNTs is synthesis by plasma rotating arc discharge technique. The centrifugal force caused by the rotation generates turbulence and accelerates the carbon vapour perpendicular to the anode. In addition, the rotation distributes the micro discharges uniformly and generates a stable plasma.

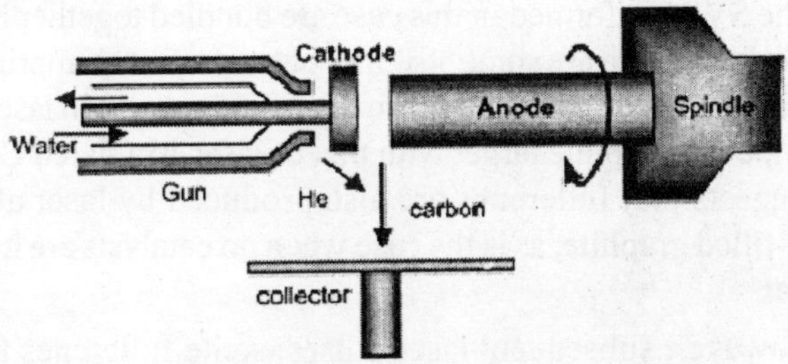

Figure 4: Schematic diagram of plasma rotating electrode system.

Consequently, it increases the plasma volume and raises the plasma temperature. At a rotation speed of 5000 rpm a yield of 60 % was found at a formation temperature of 1025°C without the use of a catalyst. The yield increases up to 90% after purification if the rotation speed is increased and the temperature is enlarged to 1150°C. The diameter size was not mentioned in this publication.

Laser Ablation

In 1995, Smalley's groupat Rice University reported the synthesis of carbon nanotubes by laser vaporisation. A pulsed, or continuouslaser is used to vaporise a graphite target in an oven at 1200°C. The main difference between continuous and pulsed laser, is that the pulsed laser demands a much higher light intensity (100 kW/cm^2 compared with 12 kW/cm^2). The oven is filled with helium or argon gas in order to keep the pressure at 500 Torr. A very hot vapour plume forms, then expands and cools rapidly. As the vaporised species cool, small carbon molecules and atoms quickly condense to form larger clusters, possibly including fullerenes.

The catalysts also begin to condense, but more slowly at first, and attach to carbon clusters and prevent their closing into cage structures. Catalysts may even open cage structures when they attach to them. From these initial clusters, tubular molecules grow into single-wall carbon nanotubes until the catalyst particles become too large, or until conditions have cooled sufficiently that carbon no longer can diffuse through or over the surface of the catalyst particles. It is also possible that the particles become that much coated with a carbon layer that they cannot absorb more and the nanotube stops growing.

The SWNTs formed in this case are bundled together by van der Waals forces. There are some striking, but not exact similarities, in the comparison of the spectral emission of excited species in laser ablation of a composite graphite target with that of laser-irradiated C_{60} vapour. This suggests that fullerenes are also produced by laser ablation of catalyst-filled graphite, as is the case when no catalysts are included in the target.

However, subsequent laser pulses excite fullerenes to emit C_2 that adsorbs on catalyst particles and feeds SWNT growth. However, there is insufficient evidence to conclude this with certaintyLaser ablation is almost similar to arc discharge, since the optimum background gas and catalyst mix is the same as in the arc discharge process. This might be due to very similar reaction conditions needed, and the reactions probably occur with the same mechanism.

SWNT and MWNT

The condensates obtained by laser ablation are contaminated with carbon nanotubes and carbon nanoparticles. In the case of pure graphite electrodes, MWNTs would be synthesised, but uniform SWNTs could be synthesised if a mixture of graphite with Co, Ni, Fe or Y was used instead of pure graphite. SWNTs synthesised this way exist as 'ropes'. Laser vaporisation results in a higher yield for SWNT synthesis and the nanotubes have better properties and a narrower size distribution than SWNTs produced by arc-discharge.

Nanotubes produced by laser ablation are purer than those produced in the arc discharge process. The Ni/Y mixture catalyst (Ni/Y is 4.2/1) gave the best yield. The size of the SWNTs ranges from 1-2 nm, for example the Ni/Co catalyst with a pulsed laser at 1470 °C gives SWNTs with a diameter of 1.3-1.4 nm. In case of a continuous laser at 1200 °C and Ni/Y catalyst (Ni/Y is 2:0.5 at. %), SWNTs with an average diameter of 1.4 nm were formed with 20-30 % yield.

Large Scale Synthesis of SWNT

Because of the good quality of nanotubes produced by this method, scientists are trying to scale up laser ablation. However the results are not yet as good as for the arc-discharge method, but they are still promising. Two of the newest developments on large-scale synthesis of SWNTs will be discussed. The first is the 'ultra fast Pulses from a free electron laser' method, the second is 'continuous wave laser-powder' method. Scaling up is possible, but the technique is rather expensive due to the laser and the large amount of power required.

Free Electron Laser (FEL) Method

Usually the pulses in an Nd:YAG system have width of approximately 10 ns, in this FEL system the pulse width is ~ 400 fs. The repetition rate of the pulse is enormously increased from 10 Hz to 75 MHz. To give the beam the same amount of energy as the pulse in an Nd:YAG system, the pulse has to be focused. The intensity of the laser bundle behind the lens reaches ~5 x 10^{11} W/cm^2, which is about 1000 times greater than in Nd:YAG systems. A jet of preheated (1000°C) argon through a nozzle tip is situated close to the rotating graphite target, which contains the catalyst.

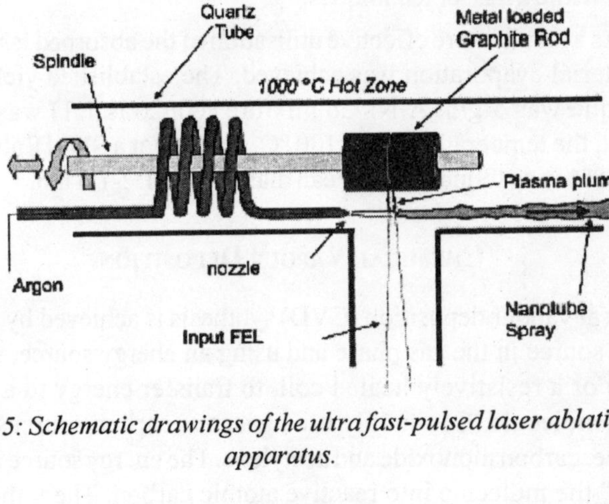

Figure 5: Schematic drawings of the ultra fast-pulsed laser ablation apparatus.

The argon gas deflects the ablation plume approximately 90° away from the incident FEL beam direction, clearing away the carbon vapour from the region in front of the target. The produced SWNT soot, is collected in a cold finger. This process can be seen in Figure 5. The yield at this moment is 1,5 g/h, which is at 20 % of the maximum power of the not yet upgraded FEL.

If the FEL is upgraded to full power and is working at 100 % power, a yield of 45 g/h could be reached since the yield was not limited by the laser power. With this method the maximum achievable yield with the current lasers is 45 g/h, with a NiCo or NiY catalyst, in argon atmosphere at 1000 °C and a wavelength of ~3000 nm. The SWNTs produced in bundles of 8-200 nm and a length of 5-20 microns has a diameter range 1-1.4 nm.

Continuous Wave Laser-powder Method

This method is a novel continuous, highly productive laser-powder method of SWNT synthesis based on the laser ablation of mixed graphite and metallic catalyst powders by a 2-kW continuous wave CO_2 laser in an argon or nitrogen stream. Because of the introduction of micron-size particle powders, thermal conductivity losses are significantly decreased compared with laser heating of the bulk solid targets in known laser techniques.

As a result, more effective utilisation of the absorbed laser power for material evaporation was achieved. The established yield of this technique was 5 g/h. A Ni/Co mixture (Ni/Co is 1:1) was used as catalyst, the temperature was 1100 °C. In the soot a SWNT abundance of 20-40% w as found with a mean diameter of 1 2-1.3 nm.

Chemical Vapour Deposition

Chemical vapour deposition (CVD) synthesis is achieved by putting a carbon source in the gas phase and using an energy source, such as a plasma or a resistively heated coil, to transfer energy to a gaseous carbon molecule. Commonly used gaseous carbon sources include methane, carbon monoxide and acetylene. The energy source is used to "crack" the molecule into reactive atomic carbon. Then, the carbon diffuses towards the substrate, which is heated and coated with a catalyst (usually a first row transition metal such as Ni, Fe or Co) where it will bind.

Carbon nanotubes will be formed if the proper parameters are maintained. Excellent alignment, as well as positional control on nanometre scale, can be achieved by using CVD. Control over the diameter, as well as the growth rate of the nanotubes can also be maintained. The appropriate metal catalyst can preferentially grow single rather than multi-walled nanotubes.

CVD carbon nanotube synthesis is essentially a two-step process consisting of a catalyst preparation step followed by the actual synthesis of the nanotube. The catalyst is generally prepared by sputtering a transition metal onto a substrate and then using either chemical etching or thermal annealing to induce catalyst particle nucleation. Thermal annealing results in cluster formation on the substrate, from which the nanotubes will grow. Ammonia may be used as the etchant. The temperatures for the synthesis of nanotubes by CVD are generally within the 650-900°C range.

Typical yields for CVD are approximately 30%. These are the basic principles of the CVD process. In the last decennia, different techniques for the carbon nanotubes synthesis with CVD have been developed, such as plasma enhanced CVD, thermal chemical CVD, alcohol catalytic CVD, vapour phase growth, aero gel-supported CVD and laserassisted CVD.

Plasma Enhanced Chemical Vapour Deposition

The plasma enhanced CVD method generates a glow discharge in a chamber or a reaction furnace by a high frequency voltage applied to both electrodes.

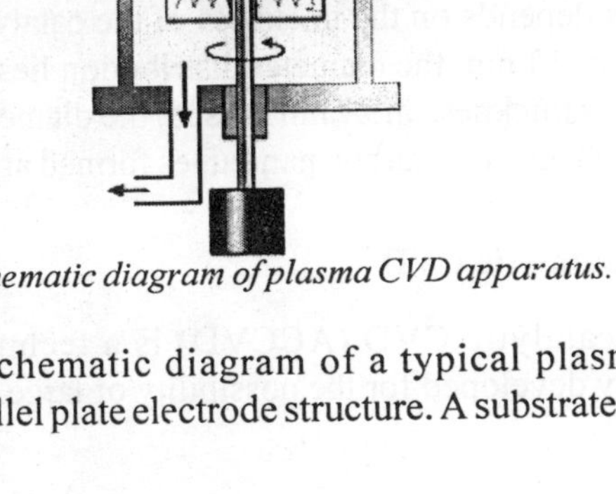

Figure 6: Schematic diagram of plasma CVD apparatus.

Figure 6 shows a schematic diagram of a typical plasma CVD apparatus with a parallel plate electrode structure. A substrate is placed on the grounded electrode. In order to form a uniform film, the reaction gas is supplied from the opposite plate. Catalytic metal, such as Fe, Ni and Co are used on for example a Si, SiO_2, or glass substrate using thermal CVD or sputtering. After nanoscopic fine metal particles are formed, carbon nanotubes will be grown on the metal particles on the substrate by glow discharge generated from high frequency power. A carbon containing reaction gas, such as C_2H_2, CH_4, C_2H_4, C_2H_6, CO is supplied to the chamber during the discharge.

The catalyst has a strong effect on the nanotube diameter, growth rate, wall thickness, morphology and microstructure. Ni seems to be the most suitable pure-metal catalyst for the growth of aligned multiwalled carbon nanotubes (MWNTs). The diameter of the MWNTs is approximately 15 nm. The highest yield of carbon nanotubes achieved was about 50% and was obtained at relatively low temperatures (below 330°C).

Thermal Chemical Vapour Deposition

In this method Fe, Ni, Co or an alloy of the three catalytic metals is initially deposited on a substrate. After the substrate is etched in a diluted HF solution with distilled water, the specimen is placed in a quartz boat. The boat is positioned in a CVD reaction furnace, and nanometre-sized catalytic metal particles are formed after an additional etching of the catalytic metal film using NH_3 gas at a temperature of 750 to 1050°C.

As carbon nanotubes are grown on these fine catalytic metal particles in CVD synthesis, forming these fine catalytic metal particles is the most important process. When growing carbon nanotubes on a Fe catalytic film by thermal CVD, the diameter range of the carbon nanotubes depends on the thickness of the catalytic film. By using a thickness of 13 nm, the diameter distribution lies between 30 and 40 nm. When a thickness of 27 nm is used, the diameter range is between 100 and 200 nm. The carbon nanotubes formed are multiwalled.

Alcohol Catalytic CVD

Alcohol catalytic CVD (ACCVD) is a technique that is being intensively developed for the possibility of large-scale production of high quality single wall nanotubes SWNTs at low cost. In this technique, evaporated alcohol, methanol and ethanol, are being utilised over iron and cobalt catalytic metal particles supported with zeolite.

Generation is possible is possible at a relatively low minimum temperature of about 550°C. It seems that hydroxyl radicals, who come from reacting alcohol on catalytic metal particles, remove carbon atoms with dangling bonds, which are obstacles in creating high-purity SWNTs. The diameter of the SWNTs is about 1 nm. The lower reaction temperature and the high-purity features of this ACCVD technique guarantee an easy possibility to scale production up at low cost.

Vapour Phase Growth

Vapour phase growth is a synthesis method of carbon nanotubes, directly supplying reaction gas and catalytic metal in the chamber without a substrate..39. Two furnaces are placed in the reaction chamber. Ferrocene is used as catalyst. In the first furnace, vaporisation of catalytic carbon is maintained at a relatively low temperature.

Fine catalytic particles are formed and when they reach the second furnace, decomposed carbons are absorbed and diffused to the catalytic metal particles. Here, they are synthesised as carbon nanotubes. The diameter of the carbon nanotubes by using vapour phase growth are in the range of 2 - 4 nm for SWNTs and between 70 and 100 nm for MWNTs.

Aero gel-supported CVD

In this method SWNTs are synthesised by disintegration of carbon monoxide on an aero gel-supported Fe/Mo catalyst. There are many important factors that affect the yield and quality of SWNTs, including the surface area of the supporting material, reaction temperature and feeding gas. Because of the high surface area, the porosity and ultra-light density of the aero gels, the productivity of the catalyst is much higher than in other methods.

After a simple acidic treatment and a oxidation process high purity (>99%) SWNTs can be obtained. When using CO as feeding gas the yield of the nanotubes is lower but the overall purity of the materials is very good. The diameter distribution of de nanotubes is between 1,0 nm and 1,5 nm. The optimal reaction temperature is approximately 860°C.

Laser-assisted Thermal CVD

In laser-assisted thermal CVD (LCVD) a medium power, continuous wave CO_2 laser, which was perpendicularly directed onto a substrate, pyrolyses sensitised mixtures of $Fe(CO)_5$ vapour and acetylene in a flow reactor. The carbon nanotubes are formed by the catalysing action of the very small iron particles. By using a reactant gas mixture of iron pentacarbonyl vapour, ethylene and acetylene both single- and multi-walled carbon nanotubes are produced. Silica is used as substrate. The diameters of the SWNTs range from 0.7 to 2.5 nm. The diameter range of the MWNTs is 30 to 80 nm.

CoMoCat Process

In this method, SWNTs are grown by CO disproportionation at 700 - 950°C. The technique is based on a unique Co-Mo catalyst formulation that inhibits the sintering of Co particles and therefore inhibits the formation of undesired forms of carbon that lower the selectivity. During the SWNT reaction, cobalt is progressively reduced from the oxidic state to the metallic form. Simultaneously Molybdenum is converted to the carbidic form (Mo_2C).

Co acts as the active species in the activation of CO, while the role of the Mo is possibly dual. It would stabilise Co as a well-dispersed Co^{2+} avoiding its reduction and would act as a carbon sink to moderate the growth of carbon inhibiting the formation of undesirable forms of carbon. It is found that one of the critical conditions for an effective reactor operation is that the space velocity has to be high enough to keep the CO conversion as low as possible.

The most important advantage of fluidised bed reactors is that they permit continuous addition and removal of solid particles from the reactor, without stopping the operation. This method can be scaled-up without losses in SWNT quality.

Table 1: Diameter range versus temperature

Temperature (°C)	Diameter range
750	0,9 ± 0,05
850	0,9 – 1,25
950	1,00 – 1,40

By varying the operation conditions, SWNTs can be produced with different diameter ranges. In Table 1 several diameter ranges at different temperatures are given.

High Pressure CO Disproportionation

The High pressure CO disproportionation process (HiPco) is a technique for catalytic production of SWNTs in a continuous-flow gas phase using CO as the carbon feedstock and $Fe(CO)_5$ as the ironcontaining catalyst precursor. SWNTs are produced by flowing CO, mixed with a small amount of $Fe(CO)_5$ through a heated reactor. Size and diameter distribution of the nanotubes can be roughly selected by controlling the pressure of CO. This process is promising for bulk production of carbon nanotubes.

Nanotubes as small as 0.7 nm in diameter, which are expected to be the smallest achievable chemically stable SWNTs, have been produced by this method. The average diameter of HiPco SWNTs is approximately 1.1 nm. The yield that could be achieved is approximately 70%. The highest yields and narrowest tubes can be produced at the highest accessible temperature and pressure. SWNT material with 97% purity can be produced at rates of up to 450 mg/h with this process.

Flame Synthesis

This method is based on the synthesis of SWNTs in a controlled flame environment, that produces the temperature, forms the carbon atoms from the inexpensive hydrocarbon fuels and forms small aerosol metal catalyst islands. SWNTs are grown on these metal islands in the same manner as in laser ablation and arc discharge. These metal catalyst islands can be made in three ways. The metal catalyst (cobalt) can either be coated on a mesh, on which metal islands resembling droplets were formed by physical vapour deposition. These small islands become aerosol after exposure to a flame.

The second way, is to create aerosol small metal particles by burning a filter paper that is rinsed with a metal-ion solution. The third way, is the thermal evaporating technique in which metal powder is inserted in a trough and heated. In a controlled way a fuel gas is partially burned to gain the right temperature of ~800°C and the carbon atoms for SWNT production. On the small metal particles the SWNTs are than formed. As optimization parameters the fuel gas composition, catalyst, catalyst carrier surface and temperature can be controlled. In the literature found, the yield, typical length and diameters are not stated.

Purification of SWNTs

A large problem with nanotube application is next to large-scale synthesis also the purification. The main impurities in the soot are graphite (wrapped up) sheets, amorphous carbon, metal catalyst and the smaller fullerenes. These impurities will interfere with most of the desired properties of the SWNTs. Also in the fundamental research, it is preferred to obtain SWNTs or the impurities, as pure as possible without changing them. In order to understand the measurements better, the SWNT samples also have to be as homogeneous as possible.

The common industrial techniques use strong oxidation and acid refluxing techniques, which have an effect on the structure of the tubes. Basically, these techniques can be divided into two mainstreams, structure selective and size selective separations. The first one will separate the SWNTs from the impurities; the second one will give a more homogeneous diameter or size distribution.

Techniques

Oxidation

Oxidative treatment of the SWNTs is a good way to remove carbonaceous impuritiesor to clear the metal surface. The main disadvantages of oxidation are that not only the impurities are oxidised, but also the SWNTs. Luckily the damage to SWNTs is less than the damage to the impurities. These impurities have relatively more defects or a more open structure. Another reason why impurity oxidation is preferred, is that these impurities are most commonly attached to the metal catalyst, which also acts as oxidising catalyst.

Altogether, the efficiency and the yield of the procedure are highly dependable on a lot of factors, such as metal content, oxidation time, environment, oxidising agent and temperature. The fact that metal acts as oxidising catalyst, the metal content should certainly be taken into consideration, when looking at the oxidising time. For example, when the temperature is raised above 600 °C, SWNTs will also oxidise, even without catalyst. This is the case with thermal, fixed airand pure oxygen oxidations. These can easily oxidise all the components, so the temperature and the time should be in good control.

There are a couple of examples for clearing the metal surface, to prepare the sample for a metal removal step. The first one is (mild) oxidising in a wet environment with soluble oxidising agents, such as H_2O_2 and H_2SO_4. These will only oxidise the defects and will clear the surface of the metal. Most commonly, the metal catalyst stays intact during these processes, but when oxygen is used in a wet atmosphere, the outer layer of the metal will be oxidised. Then, the density of this surface increases and the surface covering carbon layer ruptures. Not only the carbon impurities are oxidised now, but also the metal is partially oxidised and exposed.

Acid treatment

In general the acid treatment will remove the metal catalyst. First of all, the surface of the metal must be exposed by oxidation or sonication. The metal catalyst is then exposed to acid and solvated. The SWNTs remain in suspended form. When using a treatment in HNO_3, the acid only has an effect on the metal catalyst. It has no effect on the SWNTs and other carbon particles. If a treatment in HCl is used, the acid has also a little effect on the SWNTs and other carbon particles. The mild acid treatment (4 M HCl reflux) is basically the same as the HNO_3 reflux, but here the metal has to be totally exposed to the acid to solvate it.

Annealing

Due to high temperatures (873 - 1873 K) the nanotubes will be rearranged and defects will be consumed. The high temperature also causes the graphitic carbon and the short fullerenes to pyrolyse. When using high temperature vacuum treatment (1873 K) the metal will be melted and can also be removed.

Ultrasonication

In this technique particles are separated due to ultrasonic vibrations. Agglomerates of different nanoparticles will be forced to vibrate and will become more dispersed. The separation of the particles is highly dependable on the surfactant, solvent and reagent used. The solvent influences the stability of the dispersed tubes in the system. In poor solvents the SWNTs are more stable if they are still attached to the metal. But in some solvents, such as alcohol, monodispersed particles are relatively stable. When an acid is used, the purity of the SWNTs depends on the exposure time. When the tubes are exposed to the acid for a short time, only the metal solvates, but for a longer exposure time, the tubes will also be chemically cut.

Magnetic purification

In this method ferromagnetic (catalytic) particles are mechanically removed from their graphitic shells. The SWNTs suspension is mixed with inorganic nanoparticles (mainly ZrO_2 or $CaCO_3$) in an ultrasonic bath to remove the ferromagnetic particles. Then, the particles are trapped with permanent magnetic poles. After a subsequent chemical treatment, a high purity SWNT material will be obtained. This process does not require large equipment and enables the production of laboratory-sized quantities of SWNTs containing no magnetic impurities.

Micro filtration

Micro filtration is based on size or particle separation. SWNTs and a small amount of carbon nanoparticles are trapped in a filter. The other nanoparticles (catalyst metal, fullerenes and carbon nanoparticles) are passing through the filter. One way of separating fullerenes from the

SWNTs by micro filtration is to soak the as-produced SWNTs first in a CS_2 solution. The CS_2 insolubles are then trapped in a filter. The fullerenes which are solvated in the CS_2, pass through the filter.

A special form of filtration is cross flow filtration. In cross flow filtration the membrane is a hollow fibre. The membrane is permeable to the solution. The filtrate is pumped down the bore of the fibre at some head pressure from a reservoir and the major fraction of the fast flowing solution which does not permeate out the sides of the fibre is fed back into the same reservoir to be cycled through the fibre repeatedly. A fast hydrodynamic flow down the fibre bore (cross flow) sweeps the membrane surface preventing the build-up of a filter cake.

Cutting of SWNTs

Cutting of the SWNTs can either be induced chemically, mechanicallyor as a combination of these. SWNTs can be chemically cut by partially functionalising the tubes, for example with fluor. Then, the fluorated carbon will be driven off the sidewall with pyrolisation in the form of CF_4 or COF_2. This will leave behind the chemically cut nanotubes.

Mechanical cutting of the nanotubes can be induced by ball-milling. Here, the bonds will break due to the high friction between the nanoparticles and the nanotubes will be disordered. A combination of mechanical and chemical cutting of the nanotubes, is ultrasonical induced cutting in an acid solution. In this way the ultrasonic vibration will give the nanotubes sufficient energy to leave the catalyst surface. Then, in combination with acid the nanotubes will rupture at the defect sites.

Functionalisation

Functionalisation is based on making SWNTs more soluble than the impurities by attaching other groups to the tubes. Now it is easy to separate them from insoluble impurities, such as metal, with filtration. Another functionalisation technique also leaves the SWNT structure intact and makes them soluble for chromatographic size separation. For recovery of the purified SWNTs, the functional groups can be simply removed by thermal treatment, such as annealing.

Chromatography

This technique is mainly used to separate small quantities of SWNTs into fractions with small length and diameter distribution. The SWNTs are run over a column with a porous material, through which the SWNTs will flow. The columns used are GPC (Gel Permeation Chromatography) and HPLC-SEC (High Performance Liquid Chromatography—Size Exclusion Chromatography) columns.

The number of pores the SWNTs will flow through, depends on their size. This means that, the smaller the molecule, the longer the pathway to the end of the column will be and that the larger molecules will come off first. The pore size will control what size distribution can be separated. However, a problem is that the SWNTs have to be either dispersedor solvated. This can be done by ultrasonicationor functionalisation with soluble groups.

Potential Applications of CNTs

Energy Storage

Graphite, carbonaceous materials and carbon fibre electrodes are commonly used in fuel cells, batteries and other electrochemical applications. Advantages of considering nanotubes for energy storage are their small dimensions, smooth surface topology and perfect surface specificity. The efficiency of fuel cells is determined by the electron transfer rate at the carbon electrodes, which is the fastest on nanotubes following ideal Nernstian behaviour.

Hydrogen storage

The advantage of hydrogen as energy source is that its combustion product is water. In addition, hydrogen can be easily regenerated. For this reason, a suitable hydrogen storage system is necessary, satisfying a combination of both volume and weight limitations. The two commonly used means to store hydrogen are gas phase and electrochemical adsorption. Because of their cylindrical and hollow geometry, and nanometre-scale diameters, it has been predicted that carbon nanotubes can store a liquid or a gas in the inner cores through a capillary effect.

As a threshold for economical storage, the Department of Energy has set storage requirements of 6.5 % by weight as the minimum level for hydrogen fuel cells. It is reported that SWNTs were able to meet and sometimes exceed this level by using gas phase adsorption (physisorption). Yet, most experimental reports of high storage capacities are rather controversial so that it is difficult to assess the applications potential. What lacks, is a detailed understanding of the hydrogen storage mechanism and the effect of materials processing on this mechanism. Another possibility for hydrogen storage is electrochemical storage. In this case not a hydrogen molecule but an H atom is adsorbed. This is called chemisorption.

Lithium intercalation

The basic principle of rechargeable lithium batteries is electrochemical intercalation and deintercalation of lithium in both electrodes. An ideal battery has a high-energy capacity, fast charging time and a long cycle time. The capacity is determined by the lithium saturation concentration of the electrode materials. For Li, this is the highest in nanotubes if all the interstitial sites (inter-shell van der Waals spaces, inter-tube channels and inner cores) are accessible for Li intercalation.

SWNTs have shown to possess both highly reversible and irreversible capacities. Because of the large observed voltage hysteresis, Li-intercalation in nanotubes is still unsuitable for battery application. This feature can potentially be reduced or eliminated by processing, i.e. cutting, the nanotubes to short segments.

Electrochemical supercapacitors

Supercapacitors have a high capacitance and potentially applicable in electronic devices. Typically, they are comprised two electrodes separated by an insulating material that is ionically conducting in electrochemical devices. The capacity of an electrochemical supercap inversely depends on the separation between the charge on the electrode and the counter charge in the electrolyte.

Because this separation is about a nanometre for nanotubes in electrodes, very large capacities result from the high nanotube surface area accessible to the electrolyte. In this way, a large amount of charge injection occurs if only a small voltage is applied. This charge injection is used for energy storage in nanotube supercapacitors. Generally speaking, there is most interest in the double-layer supercapacitors and redox supercapacitors with different charge-storage modes.

Molecular Electronics with CNTs

Field emitting devices

If a solid is subjected to a sufficiently high electric field, electrons near the Fermi level can be extracted from the solid by tunnelling through the surface potential barrier. This emission current depends on the strength of the local electric field at the emission surface and its work function (which denotes the energy necessary to extract an electron from its highest bounded state into the vacuum level).

The applied electric field must be very high in order to extract an electron. This condition is fulfilled for carbon nanotubes, because their elongated shape ensures a very large field amplification. For technological applications, the emissive material should have a low threshold emission field and large stability at high current density. Furthermore, an ideal emitter is required to have a nanometre size diameter, a structural integrity, a high electrical conductivity, a small energy spread and a large chemical stability. Carbon nanotubes possess all these properties. However, a bottleneck in the use of nanotubes for applications is the dependence of the conductivity and emission stability of the nanotubes on the fabrication process and synthesis conditions.

Transistors

The field-effect transistor—a three-terminal switching device—can be constructed of only one semiconducting SWNT. By applying a voltage to a gate electrode, the nanotube can be switched from a conducting to an insulating state. Such carbon nanotube transistors can be coupled together, working as a logical switch, which is the basic component of computers.

Nanoprobes and Sensors

Because of their flexibility, nanotubes can also be used in scanning probe instruments. Since MWNTtips are conducting, they can be used in STM and AFM instruments. Advantages are the improved resolution in comparison with conventional Si or metal tips and the tips do not suffer from crashes with the surfaces because of their high elasticity.

However, nanotube vibration, due to their large length, will remain an important issue until shorter nanotubes can be grown controllably. Nanotube tips can be modified chemically by attachment of functional groups. Because of this, nanotubes can be used as molecular probes, with potential applications in chemistry and biology.

Other applications are the following:

— A pair of nanotubes can be used as tweezers to move nanoscale structures on surfaces.
— Sheets of SWNTs can be used as electromechanical actuators, mimicking the actuator mechanism present in natural muscles.
— SWNTs may be used as miniaturised chemical sensors. On exposure to environments, which contain NO_2, NH_3 or O_2, the electrical resistance changes.

Composite Materials

Because of the stiffness of carbon nanotubes, they are ideal candidates for structural applications. For example, they may be used as reinforcements in high strength, low weight, and high performance composites. Theoretically, SWNTs could have a Young's Modulus of 1 TPa. MWNTs are weaker because the individual cylinders slide with respect to each other. Ropes of SWNTs are also less strong. The individual tubes can pull out by shearing and at last the whole rope will break. This happens at stresses far below the tensile strength of individual nanotubes.

Nanotubes also sustain large strains in tension without showing signs of fracture. In other directions, nanotubes are highly flexible. One of the most important applications of nanotubes based on their properties will be as reinforcements in composite materials. However, there have not been many successful experiments that show that nanotubes are better fillers than the traditionally used carbon fibres.

The main problem is to create a good interface between nanotubes and the polymer matrix, as nanotubes are very smooth and have a small diameter, which is nearly the same as that of a polymer chain. Secondly, nanotube aggregates, which are very common, behave different to loads than individual nanotubes do. Limiting factors for good load transfer could be sliding of cylinders in MWNTs and shearing of tubes in SWNT ropes. To solve this problem the aggregates need to be broken up and dispersed or cross-linked to prevent slippage.

A main advantage of using nanotubes for structural polymer composites is that nanotube reinforcements will increase the toughness of the composites by absorbing energy during their highly flexible elastic behaviour. Other advantages are the low density of the nanotubes, an increased electrical conduction and better performance during compressive load.

Another possibility, which is an example of a non-structural application, is filling of photoactive polymers with nanotubes. PPV (Poly-p-phenylenevinylene) filled with MWNTs and SWNTs is a composite, which has been used for several experiments. These composites show a large increase in conductivity with only a little loss in photoluminescence and electro-luminescence yields. Another benefit is that the composite is more robust than the pure polymer.

Of course, nanotube-polymer composites could be used also in other areas. For instance, they could be used in the biochemical field as membranes for molecular separations or for osteointegration (growth of bone cells). However, these areas are less explored. The most important thing we have to know about nanotubes for efficient use of them as reinforcing fibres is knowledge on how to manipulate the surfaces chemically to enhance interfacial behaviour between the individual nanotubes and the matrix material.

Templates

Because of the small channels, strong capillary forces exist in nanotubes. These forces are strong enough to hold gases and fluids in nanotubes. In this way, it may be possible to fill the cavities of the nanotubes to create nanowires. The critical issue here is the wetting characteristic of nanotubes. Because of their smaller pore sizes, filling of SWNTs is more difficult than filling of MWNTs.

If it becomes possible to keep fluids inside nanotubes, it could also be possible to perform chemical reactions inside their cavities. Especially organic solvents wet nanotubes easily. In this case we could speak of a nanoreactor. One of the problems in these cases is that nanotubes are normally closed. For the latter applications we have to open the nanotubes. This is possible through a simple chemical reaction: oxidation. The pentagons in the end cap of the nanotubes are more reactive than the sidewall. So, during oxidation, the caps are easily removed while the sidewall stays intact.

ENERGY STORAGE

Two elements that can be electrochemically stored in CNTs are hydrogen and lithium. Hydrogen can also be stored in CNTs by gas phase intercalation. Three units are commonly used to describe the hydrogen and lithium contents of storage materials with:

1. [H/C] ([Li/C]) as the ratio of hydrogen (lithium) atoms per atom of storage material, in this case carbon;
2. [wt%] as the ratio of the mass of hydrogen (lithium) to the mass of storage material (the gravimetric density);
3. [kgH_2m^{-3}] as the ratio of the mass of molecular hydrogen to the volume of storage material (volumetric density).

Electrochemical Storage of Hydrogen

Experimental studies

There are two methods to store hydrogen atoms reversibly in CNTs. The second method described electrochemical charge-discharge process, in which the hydrogen absorption is controlled by the potential. The hydrogen storage capacity of the CNT samples is analysed by means of electrochemical galvanostatic measurement in a 6 M KOH electrolyte. There are commonly three electrodes in the setup: a work electrode (negative), often made of a mixture of gold or nickel with the nanotube material pressed into a pellet, a reference electrode (Hg/HgO/OH-) and a counter electrode, usually made of nickel.

During the charging process, the water in the electrolyte dissociates at the work electrode. The adsorbed atomic hydrogen may be absorbed or intercalate in the electrode or recombine at the surface to molecular hydrogen and diffuse into the electrode or form gas bubbles at the surface of the electrode.

During the discharge process, the hydrogen in the electrode recombines with the OH- ions in the electrolyte to form water molecules. The amount of hydrogen desorbed from the electrode can therefore be measured by measuring the electric charge, which is equal to the product of current and time in a galvanostatic setup. The following equation gives the reaction in the working electrode:

$$CNT + xH_2O + xe^- \underset{oxidation}{\overset{reduction}{\Longleftrightarrow}} CNT + xH^+ + xOH^- \tag{1}$$

The reaction at the counter electrode, in the case of a Ni-electrode is given by:

$$Ni(OH)_2 \underset{oxidation}{\overset{reduction}{\Longleftrightarrow}} NiOOH + H^+ + e^- \tag{2}$$

When measuring the electrochemical hydrogen storage capacity of a CNT sample, a double-layer capacitance is measured too. This effect has to be taken into account when looking at the experimental results of the capacity measurements. Measurement methods are charge-discharge measurements, cyclic voltammetry, constant current measurements, electrochemical impedance measurements and chrono amperometry.

The charge/discharge experiments are commonly performed at a constant current. During the discharge, the cell resistance leads to an overpotential. With cyclic voltammetry, different voltages are applied at a certain sweep rate and the current is measured. From the thus obtained oxidation and reduction peaks it can be speculated if the hydrogen adsorption is due to the CNTs or not. With chrono amperometry the current behaviour is studied after applying a voltage step.

From impedance spectra, performed by applying sinusoidal voltage over a certain frequency range, a 'complex' resistance of the CNT-sample can be calculated. These spectra commonly consist of two semicircles and a slope, similar to a metal hydride. The smaller semicircle in the high frequency region is probably due to the double-layer capacitance. The semicircle in the low frequency region is attributed to an electrochemical reaction.

Gao et al. have studied these spectra of CNTs decorated with metallic nickel as a function of cycling and found that the smaller semicircle hardly changed upon cycling, while the larger semicircle remarkably increased, indicating an increase in surface reaction resistance upon cyling. Several experiments have been performed in this research area with often differing conclusions.

Nutzenadel et al. demonstrated in 1999 that SWNTs could electrochemically store hydrogen, in which SWNT soot containing a few percent of 0.7 - 1.2 nm diameter SWNTs was mixed with gold as a compacting powder in a 1:4 ratio to form electrodes. Gold was used because it is noble and does not participate in any electrochemical reaction. In the nanotubes-sample, nickel, iron and C_{60} were present too.

Pure samples of both nickel and iron did not hydride during cycling and C_{60} only showed a low capacity of 14 mAh/g. For the nanotubes-sample they found a maximum capacity of 110 mAh/g, indicating that the nanotubes are responsible for the hydrogen uptake. This result corresponds to ~0.39 wt% hydrogen. Only slight capacity loss was observed after many cycles.

Dai et al. investigated the electrochemical hydrogen storage behaviour of ropes of aligned SWNTs (up to 100 mm in length and 50 mm in diameter) with a large sample quantity of 200 mg (with a purity in the range 60-70%) at 298 K under normal atmosphere. The work electrode was made without compacting powder by pressing the sample to a sheet of nickel foam.

With a constant current density of 25 mA/g, the plateau of the discharge potential was observed around -0.6 V (vs Hg/HgO) and a discharge capacity of 503 mAh/g was obtained, corresponding to a hydrogen storage capacity of 1.84 wt% in SWNTs. They observed the same maximum discharge capacity for three different samples, so the results appeared reproducible. After 100 cycles the electrode still retained 80% of the maximal capacity. The loss is believed to be due to the mechanical instability of the electrode. The investigators explain the higher capacity as due to the purity and larger mean diameter of the SWNTs.

Furthermore, no compacting powder was used, resulting in a higher percentage of SWNTs in unit volume in the work electrode. Fazle Kibria et al. compared the electrochemical hydrogen storage behaviours of CNTs grown by chemical vapour deposition (CVD), arc-discharge (AD) and laser ablation (LA). The latter were purified (90 wt%). The charge capacities of the electrodes remained unchanged with increasing cycle number, but the reversible capacity behaviour decreased with cycling. The samples with the LA grown CNTs stored the highest amount of hydrogen, 1.6 wt% (corresponding to 400 mAh/g), which was 16 times that of CVD and AD grown CNTs.

During purification, besides the elimination of metals and amorphous carbon, CNTs may separate partially from their bundles, resulting in an increase in active sites for hydrogen storage for the LA grown CNTs. They also investigated the charge-discharge behaviours for Li-, Na- and K-doped sample electrodes. Each sample obtained a higher hydrogen capacity. The Li-doped CVD electrode showed a unique discharge behaviour during cycling. With increasing cycle number, the sample showed a large capacity reversibility.

The investigators concluded from cyclic voltammetric measurements that the high hydrogen storage capacity of the alkali metal-doped CNTs originated from an increase in hydrogen adsorption sites. This is suggested to be due to the introducing of the metals in the CNT bundles and separation of tubes, but not from their chemical effect. More recently, Zuttel et al. investigated several nanotubes samples at room temperature, fabricated of a mixture of 10 mg of SWNT material with 90 mg of gold powder to form a mechanically stable and conducting electrode.

The shape of the curves is different than the shape of an equilibrium curve for a metal hydride, in which there is a wide plateau due to a phase transition occurring at a specific potential. In nanotubes, there are no well-defined interstitial sites and no phase transition is expected. A few SWNT samples showed hydrogen storage capacities in the range of 0.7-0.9 wt%. The fact that this reaction was reversible is a good indication that the measured capacity is due to hydrogen storage.

An additional discharge from the oxidation of impurities in the nanotubes would not necessarily be reversible. The electrochemically measured discharge capacity at 293 K of the nanotube samples correlated with the surface area. Only the round markers in the graph represent the surface area of CNT-samples, the fitted line corresponds to this data (the data from gas phase experiments, indicated by triangular markers, were found by Nijkamp et al. and are not included in this fit). The line intercepts the axis at the origin and the slope is 1.5 wt%/1000 m^2/g.

Theoretically, the maximum specific surface area is 2630 m^2g^{-1}. From this, they extrapolated a maximum discharge capacity of 2 wt%. Less encouraging results are from Frackowiak et al. who investigated the accumulation of hydrogen by the electrochemical decomposition of aqueous alkaline medium on a negatively polarised carbon electrode under ambient conditions.

For several SWNTs the reversible hydrogen storage capacity never exceeded 0.5 wt%. They compared several published data and found no systematic relationship between the indicated purity and the maximum discharge capacity, suggesting that SWNTs are probably not responsible for the values of sorption observed. Their final conclusion is that hydrogen storage capacity of SWNTs under a high pressure at room temperature is negligible.

References

Levins CG, Schafmeister CE., "The synthesis of curved and linear structures from a minimal set of monomers", *Journal of Organic Chemistry*, 70, p. 9002, 2005.

Lynn E., *Foster: Nanotechnology: Science, Innovation, and Opportunity*, Prentice Hall, December 21 2005.

Michael Rieth and Wolfram Schommers, *Handbook of Theoretical and Computational Nanotechnology* (10-Volume Set), American Scientific Publishers, 2006.

Narayan RJ, Kumta PN, Sfeir C, Lee D-H, Olton D, Choi D., "Nanostructured Ceramics in Medical Devices: Applications and Prospects", JOM 56 (10): 38-43, 2004.

Shetty RC., "Potential pitfalls of nanotechnology in its applications to medicine: immune incompatibility of nanodevices", *Med Hypotheses* 65 (5): 998-9, 2005.

Taniguchi, N., "On the Basic Concept of 'Nano-Technology'," Proc. Intl. Conf. Prod. London, Part II, *British Society of Precision Engineering*, 1974.

William Sims Bainbridge, *Nanoconvergence: The Unity of Nanoscience, Biotechnology, Information Technology and Cognitive Science*, Prentice Hall, June 27 2007.

7

Nanolithography

Nanolithography is used during the fabrication of leading-edge semiconductor integrated circuits or nanoelectromechanical systems (NEMS). As of 2007, nanolithography is a very active area of research in academia and in industry.

Photolithography

Photolithography (also called optical lithography) is a process used in microfabrication to selectively remove parts of a thin film. It uses light to transfer a geometric pattern from a photomask to a light-sensitive chemical on the substrate. A series of chemical treatments then engraves the exposure pattern into the material underneath the photoresist. In a complex integrated circuit, a wafer will go through the photolithographic cycle up to 50 times.

Photolithography shares some fundamental principles with photography, in that the pattern in the etching resist is created by exposing it to light, either using a projected image or an optical mask. This step is like an ultra high precision version of the method used to make printed circuit boards.

Subsequent stages in the process have more in common with etching than to lithographic printing. It is used because it affords exact control over the shape and size of the objects it creates, and because it can create patterns over an entire surface simultaneously. Its main disadvantages are that it requires a flat substrate to start with, it is not very effective at creating shapes that are not flat, and it can require extremely clean operating conditions.

Procedure of Photolithography

A single iteration of photolithography combines several steps in sequence. Modern cleanrooms use automated, robotic wafer track systems to coordinate the process. The procedure described here omits some advanced treatments, such as thinning agents or edge-bead removal.

If organic or inorganic contaminations are present on the wafer surface, they are usually removed by wet chemical treatment, e.g. the RCA clean procedure based on solutions containing hydrogen peroxide.

Preparation

The wafer is initially heated to a temperature sufficient to drive off any moisture that may be present on the wafer surface. Wafers that have been in storage must be chemically cleaned to remove contamination. A liquid or gaseous "adhesion promoter", such as Bis (trimethylsilyl) amine ("hexamethyldisilazane", HMDS), is applied to promote adhesion of the photoresist to the wafer.

The phrase "adhesion promotor" is in fact incorrect, as the surface layer of Silicondioxide on the wafer reacts with the agent to form Methylated Silicon-hydroxide, a highly water repellent layer not unlike the layer on a car's paintcoating after a thorough wash and wax job. This water repellent layer prevents the aqueous developer to penetrate between the photoresist layer and the wafer's surface thus preventing so-called lifting of small photoresist structures in the (developing) pattern.

Photoresist Application

The wafer is covered with photoresist ("PR") by spin coating. A viscous, liquid solution of photoresist is dispensed onto the wafer, and the wafer is spun rapidly to produce a uniformly thick layer. The spin coating typically runs at 1200 to 4800 rpm for 30 to 60 seconds, and produces a layer between 0.5 and 2.5 micrometres thick. The spincoating process results in an amazingly uniform thin layer, usually with an uniformity within 5 to 10 nanometers.

This uniformity can be explained by detailed fluidmechanical modelling, which shows that essentially the resist moves much faster at the top of the layer than at the bottom, where viscous forces bind the resist to the wafer surface. Thus, the top layer of resist is quickly ejected from the wafer's edge while the bottom layer still creeps slowly radially along the wafer. In this way, any 'bump' or 'ridge' of resist is removed, leaving a very flat layer. Final thickness is also determined by the evaporation of liquid solvents from the resist.

For large substrates such as Generation 5 and above LCD Flat Panel Displays, an alternative application technology is used (Slit coating). Given the dimensions of the substrate, it is impracticable to spin it (1100mm x 1850mm and above), so photoresist is applied using a series of micro nozzles spread on a axis which moves along the flat panel. The resist used is more viscous than in traditional spin application. The photoresist-coated wafer is then "soft-baked" or "prebaked" to drive off excess solvent, typically at 90 to 100°C for 5 to 30 minutes in an oven or for 30 to 60 seconds on a hotplate. Sometimes a nitrogen atmosphere is used.

Photographic Exposure and Developing

After prebaking, the photoresist is exposed to a pattern of intense light. Optical lithography typically uses ultraviolet light. Positive photoresist, the most common type, becomes soluble in the basic developer when exposed; negative photoresist becomes insoluble in the (organic) developer. This chemical change allows some of the photoresist to be removed by a special solution, called "developer" by analogy with photographic developer. To learn more about the process of exposure and development of positive resist.

A post-exposure bake is performed before developing, typically to help reduce standing wave phenomena caused by the destructive and constructive interference patterns of the incident light. The develop chemistry is delivered on a spinner, much like photoresist. Developers originally often contained sodium hydroxide (NaOH). However, sodium is considered an extremely undesirable contaminant in MOSFET fabrication because it degrades the insulating properties of gate oxides.

Metal-ion-free developers such as tetramethylammonium hydroxide (TMAH) are now used. The resulting wafer is then "hard-baked", typically at 120 to 180°C for 20 to 30 minutes. The hard bake solidifies the remaining photoresist, to make a more durable protecting layer in future ion implantation, wet chemical etching, or plasma etching.

Etching Techniques

In etching, a liquid ("wet") or plasma ("dry") chemical agent removes the uppermost layer of the substrate in the areas that are not protected by photoresist. In semiconductor fabrication, dry etching techniques are generally used, as they can be made anisotropic, in order to avoid significant undercutting of the photoresist pattern. This is essential when the width of the features to be defined is similar to or less than the thickness of the material being etched.

Wet etch processes are generally isotropic in nature, which is often indispensable for microelectromechanical systems, where suspended structures must be "released" from the underlying layer. The development of low-defectivity anisotropic dry-etch process has enabled the ever-smaller features defined photolithographically in the resist to be transferred to the substrate material.

After a photoresist is no longer needed, it must be removed from the substrate. This usually requires a liquid "resist stripper", which chemically alters the resist so that it no longer adheres to the substrate. Alternatively, photoresist may be removed by a plasma containing oxygen, which oxidizes it. This process is called ashing, and resembles dry etching.

Exposure Systems

Exposure systems typically produce an image on the wafer using a photomask. The light shines through the photomask, which blocks it in some areas and lets it pass in others. Exposure systems may be classified by the optics that transfer the image from the mask to the wafer.

Contact Lithography

A contact printer, the simplest exposure system, puts a photomask in direct contact with the wafer and exposes it to a uniform light. A proximity printer puts a small gap between the photomask and wafer. In both cases, the mask covers the entire wafer, and simultaneously patterns every die. Contact printing is liable to damage both the mask and the wafer, and this was the primary reason it was abandoned for high volume production. Both contact and proximity lithography require the light intensity to be uniform across an entire wafer, and the mask to align precisely to features already on the wafer. As modern processes use increasingly large wafers, these conditions become increasingly difficult.

Research and prototyping processes often use contact lithography, because it uses inexpensive hardware and can achieve high optical resolution. The resolution is approximately the square root of the product of the wavelength and the gap distance. Hence, contact printing offers the best resolution, because its gap distance is approximately zero. In addition, nanoimprint lithography may revive interest in this familiar technique.

Projection Systems

Very-large-scale integration lithography uses projection systems. Unlike contact or proximity masks, which cover an entire wafer, projection masks show only one die. Projection exposure systems (steppers) project the mask onto the wafer many times to create the complete pattern.

Photomasks

The image for the mask originates from a computerised data file. This data file is converted to a series of polygons and written onto a square fused quartz substrate covered with a layer of chrome using a photolithographic process. A beam of electrons is used to expose the pattern defined in the data file and travels over the surface of the substrate in either a vector or raster scan manner. Where the photoresist on the mask is exposed, the chrome can be etched away, leaving a clear path for the light in the stepper/scanner systems to travel through.

Light Sources

The evolution of lithography wavelength corresponding to different light sources. It is worth noting that the same light source may be used for several technology generations. Historically, photolithography has used ultraviolet light from gas-discharge lamps using mercury, sometimes in combination with noble gases such as xenon. These lamps produce light across a broad spectrum with several strong peaks in the ultraviolet range. This spectrum is filtered to select a single spectral line, usually the "g-line" (436 nm) or "i-line" (365 nm).

More recently, lithography has moved to "deep ultraviolet", produced by excimer lasers. Krypton fluoride produces a 248-nm spectral line, and argon fluoride a 193-nm line. Generally, changing wavelength is not a trivial matter, as the method of generating the new wavelength is completely different, as well as the absorption by different materials. Air begins to absorb significantly around the 193 nm wavelength; moving to shorter wavelengths would require installing vacuum pump and purge equipment on the lithography tools (a significant expense). Furthermore, insulating materials such as silicon dioxide (SiO_2), when exposed to photons with energy greater than the band gap, release free electrons and holes which subsequently cause adverse charging.

Optical lithography can be extended to feature sizes below 50 nm using 193 nm and liquid immersion techniques. Also termed immersion lithography, this enables the use of optics with numerical apertures exceeding 1.0. The liquid used is typically ultra-pure, deionised water, which provides for a refractive index above that of the usual air gap between the lens and the wafer surface. This is continually circulated to eliminate thermally-induced distortions. Water will only allow NA's of up to ~1.4, but materials with higher refractive indices will allow the effective NA to be increased further.

Tools using 157 nm wavelength DUV in a manner similar to current exposure systems have been developed. These were once targeted to succeed 193 nm at the 65 nm feature size node but have now all but been eliminated by the introduction of immersion lithography. This was due to persistent technical problems with the 157 nm technology and economic considerations that provided strong incentives for the continued use of 193 nm technology. High-index immersion lithography is the newest extension of 193 nm lithography to be considered. In 2006, features less than 30 nm were demonstrated by IBM using this technique.

Experimental Methods

Photolithography has been defeating predictions of its demise for many years. For instance, it was predicted that features smaller than 1 micrometre could not be printed optically. Modern techniques already print features with dimensions a fraction of the wavelength of light used—an amazing optical feat. Current research is exploring new tricks in the ultraviolet regime, as well as alternatives to conventional UV, such as electron beam lithography, X-ray lithography, extreme ultraviolet lithography, ion projection lithography, and immersion lithography.

Other Nanolithography Techniques

- X-ray lithography can be extended to an optical resolution of 15 nm by using the short wavelengths of 1 nm for the illumination. This is implemented by the proximity printing approach. The technique is developed to the extent of batch processing. The extension of the method relies on Near Field X-rays in Fresnel diffraction: a clear mask feature is "demagnified" by proximity to a wafer that is set near to a "Critical Condition". This Condition determines the mask-to-wafer Gap and depends on both the size of the clear mask feature and on the wavelength. The method is simple because it requires no lenses.
- A method of pitch resolution enhancement which is gaining acceptance is double patterning. This technique increases feature density by printing new features in between pre-printed features on the same layer. It is flexible because it can be adapted for any exposure or patterning technique. The feature size is reduced by non-lithographic techniques such as etching or sidewall spacers.
- Work is in progress on an optical maskless lithography tool. This uses a digital micro-mirror array to directly manipulate reflected light without the need for an intervening mask. Throughput is inherently low, but the elimination of mask-related production

costs—which are rising exponentially with every technology generation—means that such a system might be more cost effective in the case of small production runs of state of the art circuits, such as in a research lab, where tool throughput is not a concern.

— The most common nanolithographic technique is Electron-Beam Direct-Write Lithography (EBDW), the use of a beam of electrons to produce a pattern—typically in a polymeric resist such as PMMA.

— Extreme ultraviolet lithography (EUV) is a form of optical lithography using ultrashort wavelengths (13.5 nm). It is the most popularly considered NGL technique.

— Charged-particle lithography, such as ion- or electron-projection lithographies (PREVAIL, SCALPEL, LEEPL), are also capable of very-high-resolution patterning.

— Nanoimprint lithography (NIL), and its variants, such as Step-and-Flash Imprint Lithography, LISA and LADI are promising nanopattern replication technologies. This technique can be combined with contact printing.

— Scanning probe lithography (SPL) is a promising tool for patterning at the deep nanometer-scale. For example, individual atoms may be manipulated using the tip of a scanning tunnelling microscope (STM). Dip-Pen Nanolithography (DPN) is the first commercially available SPL technology based on atomic force microscopy.

— The furthest developed NGL remains X-ray lithography which is extensible to 15 nm resolution by use of "demagnification" in the Near Field.

— Atomic Force Microscopic Nanolithography (AFM) is a chemomechanical surface patterning technique that uses an atomic force microscope.

Nanolithography uses self-assembled monolayers of spheres (typically made of polystyrene) as evaporation masks. This method has been used to fabricate arrays of gold nanodots with precisely controlled spacings. It is possible that molecular self-assembly methods will take over as the primary nanolithography approach, due to ever-increasing complexity of the top-down approaches listed above.

Self-assembly of dense lines less than 20 nm wide in large pre-patterned trenches has been demonstrated. The degree of dimension and orientation control as well as prevention of lamella merging still need to be addressed for this to be an effective patterning technique. The important issue of line edge roughness is also highlighted by this technique.

X-RAY LITHOGRAPHY

X-ray lithography is a next generation lithography that has been developed for the semiconductor industry. Batches of microprocessors have already been produced. The short wavelengths of 0.8 nm X-rays overcome diffraction limits in the resolution of otherwise competitive optical lithography. The X-rays illuminate a mask placed in proximity to a resist-coated wafer. No lenses are used, and only rudimentary collimating mirrors. The X-rays are broadband, typically from a compact synchrotron radiation source, allowing rapid exposure.

Deep X-ray lithography uses yet shorter wavelengths, about 0.1 nm with modified procedures, to fabricate deeper structures, sometimes three dimensional, with reduced resolution. The mask consists of an X-ray absorber, typically of gold or compounds of tantalum or tungsten, on a membrane that is transparent to X-rays, typically of silicon carbide or diamond. The pattern on the mask is written by direct write electron beam lithography onto a resist that is developed by conventional semiconductor processes. The membrane can be stretched for overlay accuracy.

Most X-ray lithography demonstrations have been performed by copying with image fidelity, i.e. without magnification, 1x But with the increasing need for high resolution, X-ray lithography is now performed on the Sweet Spot, using local "demagnification by bias." Dense structures are developed by multiple exposures with translation.

Many advantages accrue from the application of 3x "demagnification": the mask is more easily fabricated; the mask to wafer gap is increased; and the contrast is higher. The technique is extensible to dense 15 nm prints. The resulting printing has high contrast. X-rays generate secondary electrons as in the cases of extreme ultraviolet lithography and electron beam lithography. While the fine pattern definition is due principally to secondaries from Auger electrons with a short path length, the primary electrons will sensitise the resist over a larger region than the X-ray exposure.

While this does not affect the pattern pitch resolution (determined by wavelength and gap), the image exposure contrast (max - min) / (max + min) is reduced since the pitch is on the order of the primary photoelectron range. The sidewall roughness and slopes are influenced by these secondary electrons as they can travel few microns in the area under the absorber; depending on exposure x-ray energy. Several prints at about 20 nm have been published. Another manifestation of the photoelectron effect is exposure to X-ray generated electrons from thick gold films used for making daughter masks. Simulations suggest that photoelectron generation from the gold substrate may affect dissolution rates.

Extreme Ultraviolet Lithography

Extreme ultraviolet lithography (also known as EUV or EUVL) is a next-generation lithography technology using the 13.5 nm EUV wavelength. EUVL is a significant departure from the deep ultraviolet lithography used today. All matter absorbs EUV radiation. Hence, EUV lithography needs to take place in a vacuum. All the optical elements, including the photomask, must make use of defect-free Mo/Si multilayers which act to reflect light by means of interlayer interference; any one of these mirrors will absorb around 30% of the incident light. This limitation can be avoided in maskless interference lithography systems. However, the latter tools are restricted to producing periodic patterns only.

The pre-production EUVL systems built to date contain at least two condenser multilayer mirrors, six projection multilayer mirrors, and a multilayer object (mask). Since the optics already absorbs 96% of the available EUV light, the ideal EUV source will need to be sufficiently bright. EUV source development has focused on plasmas generated by laser or discharge pulses. The mirror responsible for collecting the light is directly exposed to the plasma and is therefore vulnerable to damage from the high-energy ions and other debris. This damage associated with the high-energy process of generating EUV radiation has precluded the successful implementation of practical EUV light sources for lithography.

The wafer throughput of an EUVL exposure tool is a critical metric for manufacturing capacity. Given that EUV is a technology requiring high vacuum, the throughput is limited (aside from the source power) by the transfer of wafers into and out of the tool chamber, to a few wafers per hour. Another aspect of the pre-production EUVL tools is the off-axis illumination (at an angle of 6 degrees) on a multilayer mask. The resulting asymmetry in the diffraction pattern causes shadowing effects which degrade the pattern fidelity. EUVL's shorter wavelength also increases flare, resulting in increased line width roughness.

EUV Exposure of Photoresist

When an EUV photon is absorbed, photoelectrons and secondary electrons are generated by ionization, much like what happens when X-rays or electron beams are absorbed by matter. These secondary electrons have energies of a few to tens of eV and travel tens of nanometers inside photoresist before initiating the desired chemical reaction. A contributing factor for this rather large distance is the fact that polymers have significant amounts of free volume. In a recent actual EUV print test, it was found 30 nm spaces could not be resolved, even though the optical resolution and the photoresist composition were not the limiting factor.

Range of low-energy electrons

EUV directly produces photoelectrons in the ~ 80-90 eV range, and these photoelectrons produce even lower energy electrons as they slow down in the resist. These electrons will therefore have a distribution of total distances travelled, which in turn correspond to varying spatial extent of chemical effects. This reaction occurs after the electron has essentially slowed to a halt, since it is easiest to capture at that point.

EUV photoresist images often require resist thicknesses roughly equal to the pitch. This is not only due to EUV absorption causing less light to reach the bottom of the resist but also to forward scattering from the secondary electrons. Since the photon absorption depth exceeds the electron escape depth, as the released electrons eventually slow down, they dissipate their energy ultimately as heat.

An EUV dose of 1 mJ/cm^2 generates an equivalent photoelectron dose of 10.9 μC/cm^2. Current demonstration doses exceed 10 mJ/cm^2, or equivalently, 109 μC/cm^2 photoelectron dose. The use of higher doses and/or reduced resist thicknesses to produce smaller features only results in increased irradiation of the layer underneath the photoresist. This adds another significant source of photoelectrons and secondary electrons which effectively reduce the image contrast.

EUVL Defects

EUV-related defects are attributed to the inherently ionizing energy of EUV radiation.

— The first issue is positive charging, due to ejection of photoelectrons freed from the top resist surface by the EUV radiation.

— A second issue is contamination deposition on the resist from ambient or outgassed hydrocarbons, which results from EUV- or electron-driven reactions.

— A third issue is etching of the resist by oxygen, argon or other ambient gases, which have been dissociated by the EUV radiation or the electrons generated by EUV.

Ambient gases in the lithography chamber may be used for purging and contamination reduction. These gases are ionized by EUV radiation, leading to plasma generation in the vicinity of exposed surfaces, resulting in damage to the multilayer optics and inadvertent exposure of the sample.

EUV defect printability

The minimum printable defect size increases with decreasing defect height. Defects with heights as low as 0.3 nm are still significant. Of course mask defects are also a known source of defects for EUVL. The mask defects comprise multilayer defects and defects buried under the multilayer as well as pattern defects. The buried defects are particularly insidious, and even 10 nm defects may be considered risky. The phase shift caused by an undetected 3 nm mask substrate flatness variation is sufficient to produce a printable defect.

Loss in reflectivity from defect edge

The edge of a defect causes the overlying multilayer to tilt, resulting in loss or reflectivity. The principle behind this is a quarter-wavelength deviation from the flat surface produces a half-wavelength optical path difference after reflection. The light that is reflected from the flat surface is 180 degrees out of phase with the light reflected from the quarter-wavelength deviation. It has been shown that even a 1 nm deviation from flatness would lead to a substantial reduction (~20%) of the image intensity.

In fact, defects of atomic scale height can still be printable if they are at least 200 nm (FWHM) in size. Like a lens, any defect which effectively produces a phase shift scatters light outside the defect region. The amount of light that is scattered can be calculated. Furthermore, the edge of a phase defect will further reduce reflectivity by more than 10% if its deviation from flatness exceeds 3 degrees, due to the deviation from the target angle of incidence of 84 degrees with respect to the surface. Even if the defect height is shallow, the edge still deforms the overlying multilayer, producing an extended region where the multilayer is sloped. The more abrupt the deformation, the narrower the defect edge extension, the greater the loss in reflectivity.

Unexpected Resolution Limits

Given that EUV is a significant reduction in wavelength compared to current lithography wavelengths, one would expect significantly better resolution. However, the resolution is ultimately determined by the interaction volume in the image recording medium, i.e., the photoresist. As noted above, the low energy electrons released by EUV could blur the original EUV image. In addition, there are statistical effects, especially for feature areas less than 1500 square nanometers.

Shot noise

The required dose sensitivity of 5 mJ/cm^2 implies only several thousand EUV photons or so accumulate in such a small area. With the natural Poisson distribution due to the random arrival times of the photons, there is an expected natural dose variation of at least a few percent 3 sigma, making the exposure process fundamentally uncontrollable for features less than about 40 nm. Increasing the dose will reduce the shot noise, but will also increase the flare dose and generate more free electrons.

The free electrons will spread out before slowing to a stop. Since the free electron density is lower than the initial photon density, the shot noise is always effectively larger than expected from just considering the EUV dose.

Proximity effect

Secondary electron blur in EUV lithography. Since the secondary electrons are the final agents inducing chemical change in the photoresist, their distribution determines the final image. However, since their distribution is naturally random, this constitutes a large source of natural dose variation that significantly reduces the image contrast and ultimately limits the resolution. In a classic experiment by Feder et al. at IBM, an erbium layer on a PMMA resist layer was exposed to X-rays.

The erbium layer absorbed the X-rays strongly, producing low energy secondary electrons. The X-rays which were not absorbed continued to penetrate into the PMMA, where they were only lightly absorbed. Upon removal of the erbium layer and subsequent PMMA development in solvent, the resist removal rate was found to be accelerated for the top 40 nm of the PMMA film, while it was much more gradual for the rest of the film. The accelerated rate was due to the secondary electron exposure, while the gradual rate was due to the X-ray absorption. This proved the secondary electron exposure range of 40 nm.

A more recent experiment was performed by Carter et al. at MIT and University of Wisconsin-Madison, where the X-ray absorber generating the electrons was beneath the PMMA resist rather than on top of it. In this case, the accelerated dissolution of PMMA started approximately 50 nm above the substrate.

Random walk of low-energy electrons. The spatial extent of the path of a low-energy secondary electron can range from zero to many times the mean free path. The significance of this secondary electron range is the appearance of a "proximity effect" for distances on the order of 50 nm or less. This causes the exposure tolerance to be reduced dramatically as feature sizes decrease below this range.

Even though features can still print below this range, the resolution is not repeatable. The difference in experimentally determined ranges above (40 nm vs. 50 nm) is already an indication of the lack of repeatability. The secondary electron exposure can also be thought of as a blur effect. The blur is generally not included in optical-only image simulations.

Photoelectron trajectories

Kotera et al. performed EUV photoelectron trajectory simulations, showing their range to be 30 nm. The spread of the energy deposition by these electrons can account for the observed line edge roughness. The top layer exposure is effectively less because electrons emitted from the surface never come back.

Efficient photoresist heating

Ritucci et al., reported on the improved thermal ablation efficiency for EUV wavelengths compared to DUV wavelengths. Since EUV exceeds the bandgap of all materials, it is more easily absorbed than longer wavelengths, and the same dose of incident energy results in more heating; even ~100 mJ/cm^2 would be hot enough to result in ablation.

The resolution of chemically amplified photoresists is determined by thermally driven acid diffusion (spreading). It is worth noting that even at the ablation dose of 100 mJ/cm^2, the shot noise for a 1 nm pixel is still significant (3s/avg = 36%), which could severaly impact a critical dimension (CD) for which the pixel is at least 5%, i.e., 20 nm or less.

EUVL Demonstrations

In 1996, a collaboration between Sandia National Laboratories, University of California at Berkeley, and Lucent Technologies, produced NMOS transistors with gate lengths from 75 nm to 180 nm.

The gate lengths were defined by EUV lithography. The device saturation current at 130 nm gate length was ~0.2 mA/um. A 100 nm gate device showed subthreshold swing of 90 mV/decade and saturated transconductance of 250 mS/mm. A commercial NMOS at the same design rule patterned by then-state-of-the-art DUV lithography showed 0.94 mA/um saturation current and 860 mS/mm saturated transconductance. The subthreshold swing in this case was less than 90 mV/decade.

In February 2008, a collaboration including IBM and AMD, based at the College of Nanoscale Science and Engineering (CNSE) in Albany, New York, used EUV lithography to pattern the first metal layer of a 45 nm node test chip. No specific details on device performance were given. However, the lithographic performance details given still indicated much to be desired:

— CD uniformity: 6.6%

— Overlay: 17.9 nm x, 15.6 nm y, possibly correctable to 6.7 nm x, 5.9 nm y

— Power: 1 W at wafer (>200 W required for high volume), with a dose of 3.75 mJ/cm^2

— Defects: 1/sq. cm.

Apparently, the CNSE EUV tool suffered from a well-known 16% flare problem. Flare effects may be difficult to separate from the secondary electron effects discussed earlier. Also in July 2008, IMEC printed ~60 nm contacts using their installed EUV tool. Doses of 12-18 mJ/cm^2 were used. In August 2008, SEMATECH demonstrated a 22 nm half-pitch using chemically-amplified photoresist. However, even at 15 mJ/cm^2, the linewidth roughness was very significant, 5-6 nm, so that even the image pitch regularity was challenged.

Commercial Viability

EUVL has been the subject of ongoing research and development by many groups. The predicted optical resolution capability has been demonstrated. However, optical resolution is not the limiting factor for EUV. The difficulties of EUV stem mainly from the dramatically higher energy of the EUV photon (92 eV for EUV light vs. 6.4 eV for 193 nm light), which underlies the difficulty of damage-free generation and control of EUV light and confining the energy absorption within materials.

Due to its lack of readiness, EUVL has also recently been supplanted by double patterning as the lithography of choice for upcoming technology generations. Samsung and IM Flash Technologies have already started using spacer double patterning for their 3X nm NAND Flash. At the same time, Intel had also presented double patterning, specifically double exposure, as a viable alternative path to its 11 nm node.

Nanoimprint Lithography

Nanoimprint lithography is a novel method of fabricating nanometer scale patterns. It is a simple nanolithography process with low cost, high throughput and high resolution. It creates patterns by mechanical deformation of imprint resist and subsequent processes. The imprint resist is typically a monomer or polymer formulation that is cured by heat or UV light during the imprinting. Adhesion between the resist and the template is controlled to allow proper release.

Nanoimprint lithography was first invented by Prof. Stephen Chou and his students. Soon after its invention, a lot of researchers developed many different variations and implementations. At this point, nanoimprint lithography has been added to the International Technology Roadmap for Semiconductors (ITRS) for the 32 and 22 nm nodes.

Processes of Nanoimprint Lithography

There are many different types of nanoimprint lithography, but two of them are most important: thermoplastic nanoimprint lithography and photo nanoimprint lithography.

Thermoplastic nanoimprint lithography

Thermoplastic nanoimprint lithography (T-NIL) is the earliest nanoimprint lithography developed by Prof. Stephen Chou's group. In a standard T-NIL process, a thin layer of imprint resist (thermoplastic polymer) is spin coated onto the sample substrate. Then the mold, which has predefined topological patterns, is brought into contact with the sample and they are pressed together under certain pressure.

When heated up above the glass transition temperature of the polymer, the pattern on the mold is pressed into the melt polymer film. After being cooled down, the mold is separated from the sample and the pattern resist is left on the substrate. A pattern transfer process (reactive ion etching, normally) can be used to transfer the pattern in the resist to the underneath substrate.

Photo nanoimprint lithography

In photo nanoimprint lithography (P-NIL), a photo(UV) curable liquid resist is applied to the sample substrate and the mold is normally made of transparent material like fused silica. After the mold and the substrate are pressed together, the resist is cured in UV light and becomes solid. After mold separation, a similar pattern transfer process can be used to transfer the pattern in resist onto the underneath material. The use of a UV-transparent mold is difficult in a vacuum, because a vacuum chuck to hold the mold would not be possible.

Electrochemical nanoimprinting

Electrochemical nanoimprinting can be achieved using a stamp made from a superionic conductor such as silver sulfide. When the stamp is contacted with metal, electrochemical etching can be carried out with an applied voltage. The electrochemical reaction generates metal ions which move from the original film into the stamp. Eventually all the metal is removed and the complementary stamp pattern is transferred to the remaining metal.

Nanoimprint Schemes

Full wafer nanoimprint

In a full wafer nanoimprint scheme, all the patterns are contained in a single nanoimprint field and will be transferred in a single imprint step. This allows a high throughput and uniformity. An at least 8 diameter full-wafer nanoimprint with high fidelity is possible. To ensure the pressure and pattern uniformities of full wafer nanoimprint processes and prolong the mold lifetime, a pressing method utilising isotropic fluid pressure, named Air Cushion Press (ACP) by its inventors, is developed and being used by commercial nanoimprint systems.

Step and repeat nanoimprint

Nanoimprint can be performed in a way similar to the step and repeat optical lithography. The imprint field (die) is typically much smaller than the full wafer nanoimprint field. The die is repeatedly imprinted to the substrate with certain step size. This scheme is good for nanoimprint mold creation. It is currently limited by the throughput, alignment and street width issues.

Applications

Nanoimprint lithography has been used to fabricate device for electrical, optical, photonic and biological applications. For electronics devices, NIL has been used to fabricate MOSFET, O-TFT, single electron memory. For optics and photonics, intensive study has been conducted in fabrication of subwavelength resonant grating filter, polarizers, waveplate, anti-reflective structures, integrated photonics circuit and plasmontic devices by NIL. sub-10 nm nanofluidic channels had been fabricated using NIL and used in DNA strenching experiment. Currently, NIL is used to shrink the size of biomolecular sorting device an order of magnitude smaller and more efficient.

Benefits of Nanoimprint Lithography

A key benefit of nanoimprint lithography is its sheer simplicity. The single greatest cost associated with chip fabrication is optical lithography tool used to print the circuit patterns. Optical lithography requires high powered excimer lasers and immense stacks of precision ground lens elements to achieve nanometer scale resolution. There is no need for complex optics or high-energy radiation sources with a nanoimprint tool. There is no need for finely tailored photoresists designed for both resolution and sensitivity at a given wavelength. The simplified requirements of the technology lead to its low cost.

Imprint lithography is inherently a three-dimensional patterning process. Imprint molds can be fabricated with multiple layers of topography stacked vertically. Resulting imprints replicate both layers with a single imprint step, which allows chip manufactures to reduce chip fabrication costs and improve product throughput. As mentioned above, the imprint material does not need to be finely tuned for high resolution and sensitivity.

A broader range of materials with varying properties are available for use with imprint lithography. The increased material variability gives chemists the freedom to design new functional materials rather than sacrificial etch resistant polymers. A functional material may be imprinted directly to form a layer in a chip with no need for pattern transfer into underlying materials. The successful implementation of a functional imprint material would result in significant cost reductions and increased throughput by eliminating many difficult chip fabrication processing steps.

Concerns

The key concerns for nanoimprint lithography are overlay, defects, template patterning and template wear.

Overlay

The current overlay 3 sigma capability is 10 nm. Overlay has a better chance with step-and-scan approaches as opposed to full-wafer imprint.

Defects

As with immersion lithography, defect control is expected to improve as the technology matures. Defects from the template with size below the post-imprint process bias can be eliminated. Other defects would require effective template cleaning and/or the use of intermediate polymer stamps.

When vacuum is not used during the imprint process, air can get trapped, resulting in bubble defects. This is because the imprint resist layer and the template or stamp features are not perfectly flat. There is an elevated risk when the intermediate or master stamp contains depressions (which are especially easy air traps), or when the imprint resist is dispensed as droplets just before imprinting, rather than pre-spun onto the substrate. Sufficient time must be allowed for the air to escape.

Template patterning

The template patterning can currently be performed by electron beam lithography or focused ion beam patterning; however at the smallest resolution, the throughput is very slow. As a result, optical patterning tools will be more helpful if they have sufficient resolution. Optical patterning tools are already in use for the manufacturing of photomasks. Other patterning techniques (including even double patterning) may also be used.

Template wear

The use of substantial pressure to not only contact but also penetrate a layer during imprinting accelerates the wear of imprint templates compared to other types of lithographic masks.

Future applications of nanoimprint lithography may involve the use of porous low-k materials. These materials are not stiff and, as part of the substrate, are readily damaged mechanically by the pressure of the imprint process.

Removal of Residual Layers

A key characteristic of nanoimprint lithography is the residual layer following the imprint process. It is preferable to have thick enough residual layers to support alignment and throughput and low defects. However, this renders the nanoimprint lithography step less critical for critical dimension (CD) control than the etch step used to remove the residual layer.

Hence, it is important to consider the residual layer removal an integrated part of the overall nanoimprint patterning process. In a sense, the residual layer etch is similar to the develop process in conventional lithography. It has been proposed to combine photolithography and nanoimprint lithography techniques in one step in order to eliminate the residual layer.

Proximity Effects

Nanoimprint proximity effect

— *Top*: Array of depressions is more quickly filled at the edge than the centre, resulting in less imprinting at the centre of the array.

— *Bottom*: The wide space between two groups of protrusions tends to be filled slower than the narrow spaces between the protrusions, resulting in the formation of holes in the unpatterned area.

Nanoimprint lithography relies on displacing polymer. This could lead to systematic effects over long distances. For example, a large, dense array of protrusions will displace significantly more polymer than an isolated protrusion.

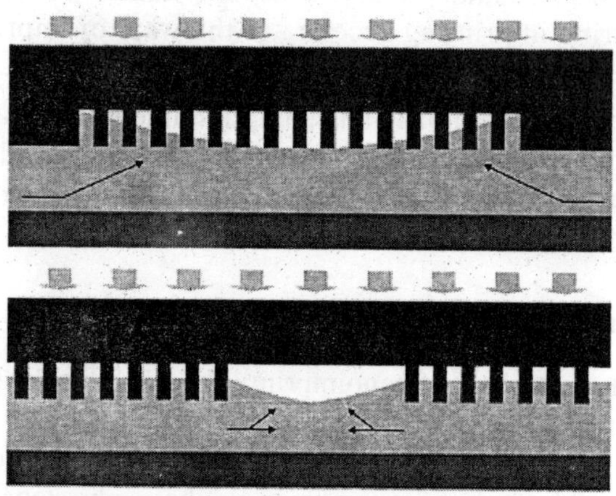

Nanoimprint proximity effect

Depending on the distance of this isolated protrusion from the array, the isolated feature may not imprint correctly due to polymer displacement and thickening. Resist holes can form in between groups of protrusions. Likewise, wider depressions in the template do not fill up with as much polymer as narrower depressions, resulting in misshapen wide lines. In addition, a depression at the edge of a large array fills up much earlier than one located in the centre of the array, resulting in within-array uniformity issues.

3D-patterning

A unique benefit of nanoimprint lithography is the ability to pattern 3D structures, such as damascene interconnects and T-gates, in fewer steps than required for conventional lithography. This is achieved by building the T-shape into the protrusion on the template.

Future of Nanoimprint

Nanoimprint lithography is a simple pattern transfer process that is neither limited by diffraction nor scattering effects nor secondary electrons, and does not require any sophisticated radiation chemistry. It is also a potentially simple and inexpensive technique. However, a lingering barrier to nanometer-scale patterning is the current reliance on other lithography techniques to generate the template.

It is possible that self-assembled structures will provide the ultimate solution for templates of periodic patterns at scales of 10 nm and less. It is also possible to resolve the template generation issue by using a programmable template in a scheme based on double patterning. As of October 2007, Toshiba is the only company to have validated nanoimprint lithography for 22 nm and beyond. What is more significant is that nanoimprint lithography is the first sub-30 nm lithography to be validated by an industrial user.

Scanning Probe Lithography

Scanning probe lithography describe a set of lithographic methods in which a microscopic or nanoscopic stylus is moved mechanically across a surface to form a pattern. This type of method can be split in two different groups :

— *Constructive* - In which the patterning is done by directly transferring chemical species to the surface

— *Destructive* - In which the patterning is done by providing the substrate with energy to physically, chemically, electronically deform the substrate's surface.

Dip-Pen Nanolithography

Dip Pen Nanolithography (DPN) is a scanning probe lithography technique where an atomic force microscope tip is used to transfer molecules to a surface via a solvent meniscus. This technique allows surface patterning on scales of under 100 nanometres. DPN is the nanotechnology analog of the dip pen, where the tip of an atomic force microscope cantilever acts as a "pen," which is coated with a chemical compound or mixture acting as an "ink," and put in contact with a substrate, the "paper."

DPN enables direct deposition of nanoscale materials onto a substrate in a flexible manner. The vehicle for deposition can include pyramidal scanning probe microscope tips, hollow tips, and even tips on thermally actuated cantilevers. Recent advances have demonstrated massively parallel patterning using two-dimensional arrays of 55,000 tips. Applications of this technology currently range through chemistry, materials science, and the life sciences, and include such work as ultra high density biological nanoarrays, additive photomask repair, and brand protection for pharmaceuticals.

The transfer of a molecular 'ink' from a coated AFM tip to a substrate was first reported by Jaschke and Butt in 1995. The technique was further developed by a research group at Northwestern University led by Chad Mirkin who also introduced the term "DPN". The company NanoInk, Inc. holds a patent on Dip Pen Nanolithography, and "DPN" and "Dip Pen Nanolithography" are trademarks or registered trademarks of NanoInk.

Soft Lithography

In technology, soft lithography refers to a family of techniques for fabricating or replicating structures using "elastomeric stamps, molds, and conformable photomasks". It is called "soft" because it uses elastomeric materials most notably PDMS. Soft lithography is generally used to construct features measured on the micrometer to nanometer scale. According to Rogers and Nuzzo, development of soft lithography expanded rapidly during the period 1995 to 2005.

Procedure

Soft lithography includes the technologies of Micro Contact Printing (μCP), replica molding (REM), microtransfer molding (μTM), micromolding in capillaries (MIMIC) and solvent-assisted micromolding (SAMIM) Patterning by etching at the nanoscale (PENs) One of the soft lithography procedures, Micro contact printing as discussed by Xia and Whitesides, is as follows:

1. The steps of any of your favourite micro- or nano- scale lithography procedures are followed to etch a desired pattern onto a substrate (usually silicon)
2. Next, the stamp is created by pouring a degassed resin overtop of the etched wafer. Common resins include PDMS and Fluorosilicone.

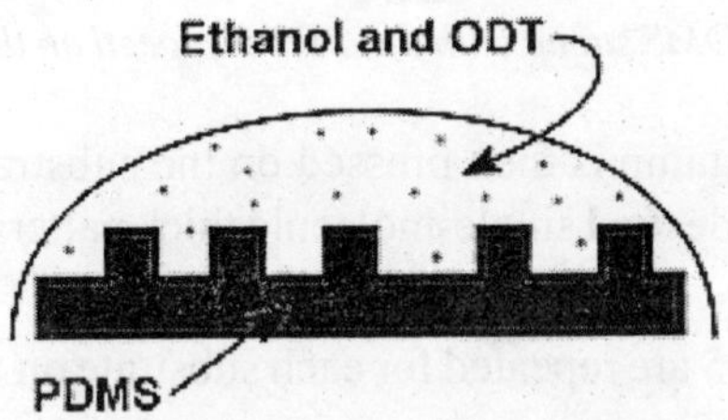

Figure 1 - "Inking" a stamp. PDMS stamp with pattern is placed in Ethanol and ODT solution

3. Removing the cured resin from the substrate, a stamp contoured to your pattern is acquired.

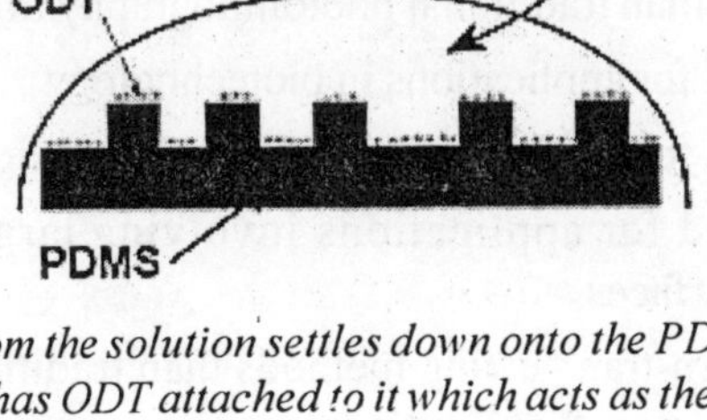

Figure 2 - ODT from the solution settles down onto the PDMS stamp. Stamp now has ODT attached to it which acts as the ink.

4. The stamp is then "inked" by placing it, pattern-up, in a bath of inking solution for a short period of time (Figure 1). The ink molecules will fall and adhere to the surface of the stamp (Figure 2) creating a single-molecule layer of the ink on the stamp.

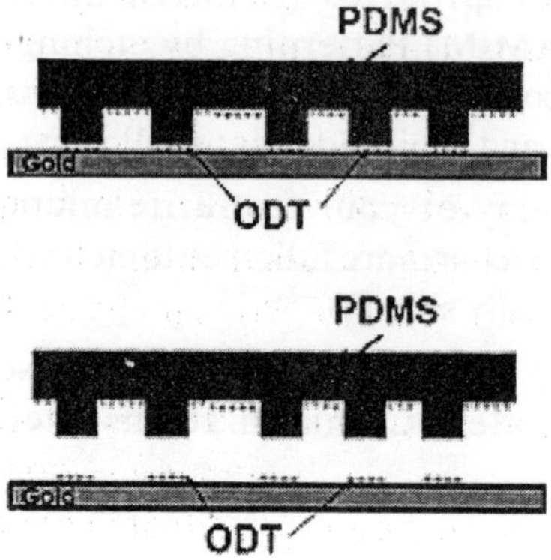

Figure 3 - The PDMS stamp with the ODT is placed on the gold substrate.

5. The inked stamp is then pressed on the substrate and removed, leaving the desired single-molecule thick pattern on the substrate (Figure 3)
6. Steps 4 and 5 are repeated for each substrate on which the pattern is desired

Advantages

Soft lithography has some unique advantages over other forms of lithography (such as photolithography and electron beam lithography). They include the following:

— Lower cost than traditional photolithography in mass production
— Well-suited for applications in biotechnology
— Well-suited for applications in plastic electronics
— Well-suited for applications involving large or nonplanar (nonflat) surfaces
— More pattern-tran·ferring methods than traditional lithography techniques (more "ink" options)
— Does not need a photo-reactive surface to create a nanostructure
— Smaller details than photolithography in laboratory settings (~30nm vs ~100nm). The resolution depends on the mask used and can reach 6 nm.

IMMERSION LITHOGRAPHY

Immersion lithography is a photolithography resolution enhancement technique that replaces the usual air gap between the final lens and the wafer surface with a liquid medium that has a refractive index greater than one. The resolution is increased by a factor equal to the refractive index of the liquid. Current immersion lithography tools use highly purified water for this liquid, achieving feature sizes below 37 nanometers. ASML Holding, Nikon and Canon are currently the only manufacturers of immersion lithography systems. An enhancement is HydroLith immersion technology which allows a "measure dry, expose wet" process.

Benefits

The lenses in the highest resolution "dry" photolithography scanners to date bend light rays through a range of angles spanning nearly 90 degrees (this angle range is known as numerical aperture). As it is impossible to increase resolution by further bending, a "last-minute" wavelength reduction is possible by inserting an immersion medium between the lens and the wafer. The wavelength is reduced by a factor equal to the refractive index of the medium. For example, for water at 193 nm wavelength, the index is 1.44.

The resolution enhancement from immersion lithography is therefore about 30-40% (depending on materials used), or about one technology node. The depth of focus, or tolerance in wafer topography flatness, is also ~2x better than a corresponding "dry" tool at the same resolution. The successful emergence of immersion lithography comes not just from its ability to extend resolution and depth of focus, but also from its timely introduction to the industry between 65 nm and 45 nm nodes.

Without immersion lithography, the semiconductor industry would have had to pursue the 45 nm node lithography using special techniques such as double patterning on dry lithography tools. Since dry optical resolution is already limited, the ability to control smaller feature sizes is more difficult. Mask costs and yield loss are naturally higher.

Manufacturing Issues

The main obstacle to adoption of immersion lithography systems has been defects and other possible sources of yield loss. Early studies focused on the elimination of bubbles in the immersion fluid, temperature and pressure variations in the immersion fluid, and immersion fluid absorption by the photoresist. Degassing the fluid, carefully constraining the fluid thermodynamics and carefully treating the top layer of photoresist have been key to the implementation of immersion lithography. Defects intrinsic to immersion lithography have been identified.

Reducing particle generation due to the water dispensing unit was found to reduce the incidence of defects. Water also has been shown to extract acid from photoresist. Specifically, photoacid generators (PAGs) are extracted into the water, which produce acid upon radiation exposure. This must be managed to ensure the lens is not corroded by the acid or contaminated by the extracted agents, and the photoresist is not chemically altered to the point of being defective.

Still, since diffusion of contaminants is expected to be much slower in water than in air or vacuum, consideration of optics contamination actually favours immersion lithography. Water-soaked photoresist also has been demonstrated to produce very satisfactory images.

The above defect concerns have led to considerations of using a topcoat layer directly on top of the photoresist. This topcoat would serve as a barrier for chemical diffusion between the liquid medium and the photoresist. In addition, the interface between the liquid and the topcoat would be optimized for watermark reduction. At the same time, defects from topcoat use should be avoided.

As scanning speeds typically approach 500 mm/s for high-volume manufacturing, the actual resist-water contact time in any given exposure area is minimal. Hence the main concerns for defects are water left behind (watermarks) and loss of resist-water adhesion (air gap). The hydrophobicity of the surface and the water delivery/removal method are therefore the key areas to address. Other areas where defects may be enhanced are at the wafer edge, where the water has to do an "about-face" (reverse motion). It is important for the water not to pick up defects from the wafer backside.

Generally, implementation into manufacturing is only considered when defect yields reach a mature level, e.g., comparable to dry lithography levels. As of 2007, many companies, including IBM, UMC, Toshiba, and TI are ramping for the 45 nm node using immersion lithography. AMD's Fab 36 is already equipped for using immersion lithography for its 65 nm, 45 nm and 32 nm node technologies. AMD has also made preparations for advanced design for manufacturability (DFM), including layout regularity and double patterning at the 22 nm node, using immersion lithography.

For the 32 nm node in 2009, Intel will begin using immersion lithography as well (source). Intel has confirmed that since EUV will not be available, it will extend 193 nm immersion lithography to the 22 nm node (source). Intel has already outlined a path to use 193 nm immersion lithography for 32 nm down to 11 nm nodes. IBM has also stated that it will be using immersion lithography for the 22 nm node, since no other alternative is available at this time (source).

Enhancements necessary to extend the technology beyond the 32 nm node are currently being investigated. Such enhancements include the use of higher refractive-index materials in the final lens, immersion fluid, and photoresist, in order to improve the resolution with single patterning. Currently, the most promising high-index lens material is lutetium aluminum garnet, with a refractive index of 2.14.

High-index immersion fluids are approaching refractive index values of 1.7. These new developments allow the optical resolution to approach ~30 nm. However, it is expected that at some point below 40 nm, current photoresists will limit further scaling. Polarisation effects due to high angles of interference in the photoresist also have to be considered as features approach 40 nm. Hence, new photoresists will need to be developed for sub-40 nm applications.

On the other hand, double patterning has received interest recently since it can potentially increase the half-pitch resolution by a factor of 2. This could allow the use of immersion lithography tools beyond the 32 nm node, potentially to the 16 nm node. While double patterning improves pitch resolution, it must rely on non-lithographic methods, such as trimming, to actually reduce the feature size, possibly by as much as 50%.

INTERFERENCE LITHOGRAPHY

Interference lithography (or holographic lithography) is a technique for patterning regular arrays of fine features, without the use of complex optical systems or photomasks.

Principle

The basic principle is the same as in interferometry or holography. An interference pattern between two or more coherent light waves is set up and recorded in a recording layer (photoresist). This interference pattern consists of a periodic series of fringes representing intensity minima and maxima. Upon post-exposure photolithographic processing, a photoresist pattern corresponding to the periodic intensity pattern emerges.

For 2-beam interference, the fringe-to-fringe spacing or period is given by $(\gamma/2)/\sin(\theta/2)$, where γ is the wavelength and θ is the angle between the two interfering waves. The minimum period achievable is then half the wavelength. By using 3-beam interference, arrays with hexagonal symmetry can be generated, while with 4 beams, arrays with rectangular symmetry are generated. Hence, by superimposing different beam combinations, different patterns are made possible.

Coherence Requirements

For interference lithography to be successful, coherence requirements must be met. First, a spatially coherent light source must be used. This is effectively a point light source in combination with a collimating lens. A laser or synchrotron beam are also often used directly without additional collimation. The spatial coherence guarantees a uniform wavefront prior to beam splitting.

Second, it is preferred to use a monochromatic or temporally coherent light source. This is readily achieved with a laser but broadband sources would require a filter. The monochromatic requirement can be lifted if a diffraction grating is used as a beam splitter, since different wavelengths would diffract into different angles but eventually recombine anyway. Even in this case, spatial coherence and normal incidence would still be required.

Beam Splitter

Coherent light must be split into two or more beams prior to being recombined in order to achieve interference. Typical methods for beam splitting are Lloyd mirrors, prisms and diffraction gratings.

Electron Holographic Lithography

The technique is readily extendible to electron waves as well, as demonstrated by the practice of electron holography. Spacings of a few nanometers or even less than a nanometer have been reported using electron holograms. This is because the wavelength of an electron is always shorter than for a photon of the same energy.

The wavelength of an electron is given by the deBroglie relation h/p, where h is the Planck constant and p is the electron momentum. For example, a 1 keV electron has a wavelength of slightly less than 0.04 nm. A 5 eV electron has a wavelength of 0.55 nm. This yields X-ray-like resolution without depositing significant energy. In order to ensure against charging, it must be ensured that electrons can penetrate sufficiently to reach the conducting substrate. A fundamental concern for using low-energy electrons (<<100 eV) with this technique is their natural tendency to repel one another due to Coulomb forces as well as Fermi-Dirac statistics, though electron anti-bunching has been verified only in a single case.

Atom Holographic Lithography

The interference of atomic deBroglie waves is also possible provided one can obtain coherent beams of cooled atoms. The momentum of an atom is even larger than for electrons or photons, allowing even smaller wavelengths, per the deBroglie relation. Generally the wavelength will be smaller than the diameter of the atom itself.

Uses

The benefit of using interference lithography is the quick generation of dense features over a wide area without loss of focus. Hence, it is commonly used for testing photoresist processes for lithography techniques based on new wavelengths. Electron interference lithography may be used for patterns which normally take too long for conventional electron beam lithography to generate.

The drawback of interference lithography is that it is limited to patterning arrayed features only. Hence, for drawing arbitrarily shaped patterns, other photolithography techniques are required. In addition, non-optical effects, such as secondary electrons from ionizing radiation or photoacid generation and diffusion, cannot be avoided with interference lithography.

For instance, the secondary electron range is roughly indicated by the width of carbon contamination (~20 nm) at the surface induced by a focused (2 nm) electron beam. This indicates that the lithographic patterning of 20 nm half-pitch features or smaller will be significantly affected by factors other than the interference pattern, such as the cleanliness of the vacuum.

Next-generation Lithography

Next-Generation Lithography (NGL) is a term used in integrated circuit manufacturing to describe the lithography technologies slated to replace photolithography. Currently the most advanced form of photolithography is immersion lithography, in which water is used as an immersion medium for the final lens. It is being applied to the 45 nm and 32 nm nodes. Several companies, including IBM, Intel and TSMC, have prepared for the continued use of current lithography, using double patterning, for the 22 nm and 16 nm nodes.

Candidates for next-generation lithography include: extreme ultraviolet lithography (EUV-lithography), X-ray lithography, electron beam lithography, focused ion beam lithography, and nanoimprint lithography. Electron beam lithography was most popular during the 1970s, but was replaced in popularity by X-ray lithography during the 1980's and early 1990's, and then by EUV lithography from the mid-1990's to the mid-2000's.

Focused ion beam lithography has carved a niche for itself in the area of defect repair. Nanoimprint's popularity is rising, and is positioned to succeed EUV as the most popular choice for next-generation lithography, due to its inherent simplicity and low cost of operation as well as its success in the LED, hard-disk and microfluidics sectors.

The rise and fall in popularity of each NGL candidate largely hinged on its throughput capability and its cost of operation and implementation. Electron beam and nanoimprint lithography are limited mainly by the throughput, while EUV and X-ray lithography are limited by implementation and operation costs. The projection of charged particles (ions or electrons) through stencil masks was also popularly considered in the early 2000's but eventually fell victim to both low throughput and implementation difficulties.

Ironically, each NGL candidate faced more competition from the extension of photolithography than from any other NGL candidate, as more and more methods of improving photolithography continued to be developed, including optical proximity correction, off-axis illumination, phase-shift masks, liquid immersion lithography, and double patterning. Even within the area of photolithography, there is a list of "next-generation" techniques, including two-photon lithography, 157 nm wavelength, and high-index immersion.

NGL issues

Reduction of k1 as seen in Intel's manufacturing. The difficulty of lithography (indicated by lowering k1) has generally increased as features shrunk, requiring investments in resolution enhancement by leading manufacturers such as Intel. Beyond 32 nm, double exposure allows an evolutionary trend, while inserting EUVL would be a revolutionary change (blue dashed lines), instantly allowing lagging technology users to afford the same resolution. Regardless of whether NGL or photolithography is used, etching of polymer (resist) is the last step. Ultimately the quality (roughness) as well as resolution of this polymer etching limits the inherent resolution of the lithography technique.

The above-mentioned competition between NGL and the recurring extension of photolithography, where the latter consistently wins, may be more a strategic than a technical matter. If a highly scalable NGL technology were to become readily available, late adopters of leading-edge technology would immediately have the opportunity to leap-frog the current use of advanced but costly photolithography techniques, at the expense of the early adopters of leading-edge technology, who have been the key investors in NGL.

While this would level the playing field, it is disruptive enough to the industry landscape that the leading semiconductor companies would probably not want to see it happen.

The following example would make this clearer. Suppose company A manufactures primarily at 180 nm, while company B manufactures down to 45 nm, by extending its photolithography capability through years of successive resolution enhancements, such as wavelength change, more and more aggressive optical proximity correction, more and more costly phase-shift masks, and finally converting to immersion lithography.

If an NGL were deployed for the 32 nm node, both companies would benefit, but company A currently manufacturing at the 180 nm node would benefit much more because it would immediately be able to use the NGL foı manufacturing at all design rules from 180 nm down to 32 nm, while company B would only benefit starting at the 32 nm node, having already spent much on extending photolithography from its 180 nm process down to 45 nm.

The gap between Company B, whose customers expect it to advance the leading edge, and Company A, whose customers don't expect an equally aggressive roadmap, will continue to widen as NGL is delayed and photolithography is extended at greater and greater cost, making the deployment of NGL less and less attractive strategically for Company B.

This becomes more clear when considering that each resolution enhancement technique applied to photolithography generally extends the capability by only one or two generations. For this reason, the observation that "optical lithography will live forever" will likely hold, as the early adopters of leading-edge technology will never benefit from highly scalable lithography technologies in a competitive environment.

Quantum Lithography

Quantum lithography is a type of photolithography, which exploits non-classical properties of the photons, such as quantum entanglement, in order to achieve superior performance over ordinary classical lithography. Quantum lithography is closely related to the fields of quantum imaging, quantum metrology, and quantum sensing.

The effect exploits the quantum mechanical state of light called the NOON state.

Of particular importance, quantum lithography can beat the classical Rayleigh criterion for the diffraction limit. Classical photolithography has an optical imaging resolution that cannot be smaller than the wavelength of light used. For example, in the use of photolithography to mass produce computer chips, it is desirable to produce smaller and smaller features on the chip, which classically requires moving to smaller and smaller wavelengths (ultraviolet and x-ray), which entails exponentially greater cost to produce the optical imaging systems at these extremely short optical wavelengths.

Quantum lithography exploits the quantum entanglement between specially prepared photons in the NOON state to achieve the smaller resolution without the requirement of shorter wavelengths. For example, a beam of red photons, entangled ten at a time in the NOON state, would have the same resolving power as a beam of x-ray photons. The field of quantum lithography is in its infancy, and although experimental proofs of principal have been carried out using the Hong-Ou-Mandel effect, it is still a long way from commercial application.

References

Akhlesh Lakhtakia (ed), *The Handbook of Nanotechnology*, Nanometer Structures: Theory, Modeling, and Simulation, SPIE Press, Bellingham, WA, USA., 2004.

Buffat, Ph.; Burrel, J.-P., "Size effect on the melting temperature of gold particles", *Physical Review* A 13 (6): 2287-2298, 1976.

Choy JH, Jang ES, Won JH, Chung JH, Jang DJ, and Kim YW., "Hydrothermal route to ZnO nanocoral reefs and nanofibers", *Appl. Phys. Lett.* 84 287, 2004.

Das S, Gates AJ, Abdu HA, Rose GS, Picconatto CA, Ellenbogen JC., "Designs for Ultra-Tiny, Special-Purpose Nanoelectronic Circuits", *IEEE Transactions on Circuits and Systems* I 54 (11): 2528-2540, 2007.

Davis, R. C., *et al.*, "Chemomechanical surface patterning and functionalization of silicon surfaces using an atomic force microscope", *Appl. Phys. Lett.* 82 (5): 808-810, 2003.

Fritz Allhoff, Patrick Lin, James Moor, and John Weckert (eds.), *Nanoethics: The Ethical and Societal Implications of Nanotechnology,* Hoboken: John Wiley & Sons, 2007.

Ghalanbor Z, Marashi SA, Ranjbar B., "Nanotechnology helps medicine: nanoscale swimmers and their future applications", *Med Hypotheses* 65 (1): 198-199, 2005.

Hatzor-de Picciotto, A., A. D. Wissner-Gross, G. Lavallee, P. S. Weiss, "Arrays of Cu(2+)-complexed organic clusters grown on gold nano dots". *Journal of Experimental Nanoscience* 2: 3-11, 2007.

Sundrani D, Darling SB, Sibener SJ, "Hierarchical assembly and compliance of aligned nanoscale polymer cylinders in confinement", *Langmuir 20* (12): 5091-9, June 2004.

8

Nanotechnology in Society

The vision of nanotechnology could be described as atomic level systems engineering. Atoms, like so many Lego blocks, can be arranged one-by-one to build microscopic analogues of macroscopic machines or indeed anything one could imagine, from scratch. This is the life analog, the construction block view of both Feynman and Drexler, whom the latter calls machine-phase nanotechnology. From drugs designed from the bottom up to artificial nano-machines that could flow through and oxygenate blood on demand, the vision is dramatic.

Drexler imagines assemblers, i.e. nanomachines that build other nanomachines, able to tackle problems from the size of the body to the globe. Take wood, an undisputedly useful building material "grown" by nanoscale/microscale machines. What else could be "grown" if we design the life that grows it? The implications, like those of biotechnology, can be scary. Fortunately, the vision is hard, and the reason this is so is because reality starts to intrude.

The novel physics, surface science, and chemistry that begin to emerge in the nanoscale regime makes the construction block paradigms fall apart. The reality is actually much more interesting. Feynman himself pointed out that quantum effects and the complex behaviour of many interacting particles—be they electrons, atoms, or molecules—would make this building block strategy difficult. In essence, that difficulty is nanotechnology today. But even he couldn't have imagined how big and how interesting those complications would be…and more importantly, how useful. Many scientists could claim the authority to explain nanotechnology—biologists, chemists, materials scientists, and physicists.

Scope of Nanotechnology

There are many reports of varying quality that describe examples and applications of nanotechnology, from the perspectives of length-scale, medicine, environment, risk, society, and public policy. If the social construct of nanotechnology encompasses both the vision and the reality, one must consider the scope of both. Oddly, the vision of nanotechnology includes both the fantastic almost science-fiction visions of nano-assemblers and artificial life but also very non-nano visions including small machines and devices.

Compared to a nanometer biological "machines" can be relatively huge; a red blood corpuscle is approximately 7000 nm while bacteria is 1000 nm and a virus is typically 60-100 nm. And key body dimensions are downright enormous. Swallowing a millimeter-sized pill that can take pictures of your insides, though it might be called nanotechnology, may have no nano-fabricated components whatsoever. This imprecision is present in both perspectives. In many ways the reality of what's going on in nanotechnology today—as presently named—is broader than even the vision.

At a recent conference on quantum nanoscience in Australia, a physics professor stood up and said something representative of many scientists' viewpoint: "Can't we just call it novel-technology?". In the policy-technology world nanotechnology has become a catchall for all advanced materials technologies, from bio-integration to molecular transistors to quantum mesoscopics. The USA National Nanotechnology Initiative is an example of this, as its proponents seek to make nanotechnology a unifying theme across many of the sciences that deal with fabrication and manipulation of matter at small-scales.

Despite these facts-on-the-ground, some have claimed that atomic and molecular engineering alone is nanotechnology. It's simply too late for this narrow and somewhat boring definition. Additionally, there are convincing reasons to lump larger systems under the nanotechnology roof. All the knowledge gained through the study of mesoscopic physics and complex systems might one day help the atomic-level construction technology. The science and technology community has to deal with putting these systems together and with the complex emergent behaviour that implies. What quantum mesoscopics, for example, gives us is a physical window on nature's immense complication.

It would also be easy to say that the vision or molecular assembly view, as theorised by Drexler's so-called machine-phase nanotechnology, is the "true" nanotechnology and stop right here. But most of the money being spent on nanotechnology and most of the people who believe they are working on nanotechnology are in the reality camp. The glow that the term nanotechnology exudes not only attracts new students to science but also attracts scientists to adopt the term and become "nanotechnologists." And it helps with funding.

From a societal standpoint, the consequences of this emerging collection of new artifacts, both directly as threats to the environment and the health, and indirectly via the transformative properties of life-changing technologies, have unquestionable importance. But there are limits to the extent of nanotechnology's embrace. In some situations, e.g. federal funding or patent decisions, more precise categorisation can become necessary. For example the patent office makes a distinction that a new invention just can't be smaller to be patentable - there has to be something new there.

Many of the suggestions Feynman made just took advantage of making things smaller and were not necessarily based on new phenomena at that length scale. In the research world some inventions or lines of discovery, though they might coincidentally take place at the nanoscale, are better classified in a different scientific context. These are often one of disparate yet specific subfields. The reality of nanotechnology has matured at its own rate as technological manufacturing and measurement techniques have steadily improved over the last 50 years, putting fabrication and control of the nano realm within reach starting some 15 years ago. It's easy to realise that "seeing" was the key to progress.

Indeed, improvements in microscopes such as the electron microscope to the creation of new classes of devices such as atomic force microscopes (AFMs) and many other techniques contributed to advances in small science. Much of this is just the natural progress of science and technology, driven by human curiosity, economic interests like the semiconductor or drug industries, and luck.

Table 1: Nanotechnology circa 2006, vision versus reality, a pragmatic definition

- The vision: From the fantastic to the mundane the vision of nanotechnology began over 40 years ago and encompasses the building block view of building matter and machines, and artificial life. Programmable machines you can't see.
- *Examples:* atom-by-atom or molecule-by-molecule construction of everything from drugs to tiny robots; not necessarily nanometer-sized but also just small (e.g. cameras or surgical machines that can be swallowed); self-replicating nano-machines (or assemblers).

- *Threats:* unknown, but greatly imagined: prolonging human life indefinitely; uncontrollable, replicating nano-assemblers turning the world into grey goo.
- *Biggest questions:* Is this a joke? How will modifying or extending the human body change civilisation? What about being able to "grow" anything cheaply with minimal materials?
- *The reality:* The science and technology behind nanotechnology today predominately defined as new phenomena that emerge in the nanometer length regime in physical, chemical, and biological systems.
- *Examples:* nanoparticles that are 1) many times more reactive than microparticles, 2) have changing properties – like colour – due to quantum confinement effects, or 3) have unknown biological effects like increased absorption in the body; advanced materials technology; quantum "mesoscopic" physics; everything and anything.
- *Threats:* possible environmental hazard as nanoparticles can be much more reactive than larger particles and may linger longer; potential risk to human and animal health due to unknown effects on the body.
- *Biggest questions:* How do we regulate these materials? How do we classify them and determine their risk?
- *Nano-enabling technologies:* Fabrication and measurement techniques that aid in the realisation of the above two definitions but do not fall within their purview. The reliable manufacturing of devices and structures with at least two dimensions at the nanoscale.

The construction block view of nanotechnology is divorced from the daily interests of most scientists and so far away as to be utterly ignorable. In some sense it is a very simplistic idea. "Hey! Let's design things from the ground up!" Let's just make the nanoscopic equivalent of a macroscopic device. Replace the surgeon and scalpel with the mini-robot. It might be more difficult to do, for various reasons, but you are not necessarily using any new science to do it. Well, not so fast.

Complexity is an inherent problem in fields from computer science to biology to physics. It is the problem of the next one thousand years. Nanotechnology is not like building the first bridge. While it is clearly too soon for the vision of nanotechnology to be a wholly engineering endeavour, the beginnings are upon us.

Nanoparticles like quantum dots and carbon nanotubes are the quintessential reality of nanotechnology today. They can exhibit new nanoscale phenomena in physics, chemical, or biological manifestations. Even if the theories of quantum physics, chemistry, and biology are complete, how they manifest themselves in complex situations is not. Progress in the nanoscience of mesoscopic and biological-materials systems will be the foundation of larger artificial small systems.

Threats and Futures and Politics

Fear as much as hope has popularised nanotechnology, from the hysterical visions of "grey goo" to the quite realistic fears of human toxicity. The image of machines invisible to the naked eye wreaking havoc has certainly captured the public's imagination. But aren't self-replicating robots that take over the world bad at any scale? In truth, many of the fears of robotics and artificial intelligence that have been around for decades have just been transferred to the invisible realm of nanotechnology, adding another dimension of fear.

Some of the new phenomena that nanoscale systems exhibit can be just as profoundly harmful as useful. Particularly relevant are the chemical and biological effects. The same reactivity that makes nanoparticles powerful in military-grade explosives can also have dramatic impact on environmental systems. And nanotoxicity is an even more serious concern. Nanoparticle transmission and accumulation through the body is largely unknown. So investment into the study of environmental and health impact is necessary.

The real issue is why there has been so much funding for the societal implications of nanotechnology. The initial reason is to deal with the vision—but the fears of the vision are ridiculous at this early point in time. A reputable scientist would find it laughable. Seriously, where are the nanomachines? Stuck on a chip in a refrigerator in the basement of some university department? Funding the study of futures like this doesn't seem responsible. Biology today is much more scary. The possible societal harm of nanotechnology pales in comparison to the prospect of an avian flu epidemic, designer babies, or proliferating suitcase nukes. Even so, the funding of bioethics is currently minimal and of questionable value.

Industry is desperate not to repeat the fiascos of genetically modified organisms (GMO) in food—now banned in Europe—and to a lesser extent nuclear energy. What better way to "educate" the public then to fund a mass of public affairs and ethics professors to study the societal effects of nanotechnology? People tend to be more positive toward ideas when financially dependent on them. Let's be clear, many of the fears disseminated to the public via science fiction and overzealous technology pundits or "futurists" are irrational; here education is important and worthwhile. Nor are professors and thinkers doing anything untoward. But it is interesting to note that the calls for all-out moratoriums on nanotechnology, ridiculous though they are, come exclusively from independent non-profit groups.

When asked about nanotechnology in an interview for the San Francisco Chronicle in 2004 former Intel CEO Craig Barrett made some illustrative remarks that are worth repeating:

> "Nanotechnology is a buzzword that you in the press have popularised, and the government popularises it. The formal definition of nanotechnology is anything below 100 nanometers. Every transistor that we make is below 100 nanometers. So, if you want to know the investment that we're making in nanotechnology, it's the total investment that Intel is making.... We don't happen to be the nanotechnology that you popularise with carbon nanotubes and quantum dots and organic molecules that are going to replace CMOS transistors. Most of that is an esoteric and populist impression of what nanotechnology is. The bulk of nanotechnology is what companies like Intel and Texas Instruments and others do today."

Is Intel really the largest nanotechnology company on earth? Is everything smaller than 100 nm nanotechnology? Or is this just CEO spin? Certainly under the bipolar transition of nanotechnology above the combination of vision and reality Intel's latest and greatest CMOS transistor evolution is not nanotechnology. It's just the logical progression of microtechnology.

First of all, companies like Intel are not taking advantage of any new nanoscale phenomena. In fact, quantum effects (like leakage via electron tunnelling) are a major problem in the most advanced integrated-circuit design. Nor are they building transistors from the ground up atom-by-atom with molecules. It may be that as the biggest dimension of Intel's transistors fall below the 10 nm length scale they will have to make use of quantum size effects or switch to another paradigm—then maybe they can claim the title.

Definitions matter in contexts relevant to nanotechnology, including patenting, risk analysis, proper funding allocation, and education. Like it or not, nanotechnology is largely perceived by the public in two ways. One, as Star Trek and most science fiction conveys it: "machines you can't see"—the more fantastic vision perspective. Nanotechnology has become what we imagine it to be. From the self-cleaning walls, to the nanites of Star Trek, to the grow anything future of The Diamond Age.

Second, the public sees the toxicology issue—dangerous particles in food and cosmetic products. This is exaggerated by the wide promotion of the term as marketing tool whether justified or not. The Woodrow Wilson Center has compiled a database of all products that claim to use nanotechnology. But the companies that are actually using the nanotechnology of today "nanoparticles" in situations that may be hazardous (such as in cosmetics) do not advertise this fact.

Nanotechnology has clearly caught the public's imagination, but they are missing the great stuff—the reality. There is a complete disconnect between what the public perceives and fears and what government agencies are defining as nanotechnology. Is the public seeing more than the toxicology or nano-robot issues? The last century taught us that chemicals could do truly awful things.

Governments and corporations around the world are funding nanotechnology for the promise of the next big thing and as a unifying theme in materials science development. Understanding how society—through the policy makers, thinkers, scientists, engineers, speakers—are defining and shaping the umbrella that is nanotechnology is the goal.

DEFINITIONS AD INFINITUM

Definitions for nanotechnology abound from government programme managers and institutions, corporate entities, individual researchers, and non-profits. Let us collect and analyse some of them here. Since these are the institutions that are driving the debate and/or funding the evolution of nanotechnology, this is a worthy goal. The strongest shaper of the direction of nanotechnology is the United States federal government acting through its research funding agencies, primarily the National Science Foundation and various military equivalents.

Mihail Roco, current head of the National Nanotechnology Initiative (NNI) and cheerleader for it even before its inception under President Clinton has given a number of definitions of nanotechnology that are often quoted. Since the NNI is shaping the future of nanotechnology by the power of federal grants, this is a good place to start. Roco's definitions are:

> "Nanotechnology is the creation of functional materials, devices, and systems through control of matter on the nanometer length scale, exploiting novel phenomena and properties (physical, chemical, biological) present only at that length scale."

There is an alternate definition as well:

> "The field of nanotechnology deals with materials and systems having these key properties: they have at least one dimension of about one to 100 nanometers, they are designed through processes that exhibit fundamental control over the physical and chemical attributes of molecular-scale structures, and they can be combined to form larger structures."

And another:

> "Nanotechnology is the ability to understand, control, and manipulate matter at the level of individual atoms and molecules, as well as at the "supramolecular" level involving clusters of molecules.

Its goal is to create materials, devices, and systems with essentially new properties and functions because of their small structure."

According to Roco, the NNI definition encourages new contributions that were not possible before:

— "novel phenomena, properties and functions at nanoscale, which are nonscalable outside of the nanometer domain;

— the ability to measure / control / manipulate matter at the nanoscale in order to change those properties and functions;

— integration along length scales, and fields of application."

Obviously Roco is firmly in the reality camp of nanotechnology, as are most of the definitions to be described. Before we analyse it further let's look at some similar definitions.

Table 2: Four Generations of Nanotechnology according to NNI: Timeline for beginning of industrial prototyping and technology commercialisation

1st:	Passive nanostructures. Example: coatings, nanoparticles, nanostructured metals, polymers, ceramics
2nd:	Active nanostructures. Example: transistors, amplifiers, targeted drugs, actuators, adaptive structures
3rd:	Systems of nanosystems. Example: guided assembling, 3D networking and new hierarchical architectures, robotics, evolutionary
4th:	Molecular nanosystems. Example: molecular devices 'by design', atomic design, emerging functions

The United Kingdom, which has been ahead of the game in considering the implications and risks of nanotechnology, put together a widely cited report in 2004. The Royal Society of London report defines nanotechnology as:

"Nanoscience is the study of phenomena and manipulation of materials at atomic, molecular, and macromolecular scales, where properties differ significantly from those at a larger scale."

This definition is often quoted outside of the United States. Roco is seeking to ignite a new "man-on-the-moon" scale investment in science and technology in this country. He is looking at nanotechnology as a unifying theme to advance all advanced technologies, which are based largely on increased understanding of the sciences made in the past century. His definition is very biased and broad to encourage these opportunities. Business as well has a motivation to make nanotechnology widely used, at least for now, because "cool" sells.

Recently, the elite academic publisher Nature (London) spun off a new journal called Nature Nanotechnology. It's illustrative to find out what they consider acceptable for submission.

Table 3: Nature Nanotechnology sub-fields suitable for submission

- Nanomaterials and nanoparticles
- Carbon nanotubes and fullerenes
- Organic-inorganic nanostructures
- Structural properties
- Electronic properties and devices
- Nanomagnetism and spintronics
- Photonic structures and devices
- Quantum information
- Molecular self-assembly
- Molecular machines and nanoelectromechanical devices (NEMS)
- Surface patterning and imaging
- Nanofluidics, nanosensors and other devices
- Nanobiotechnology and nanomedicine
- Computational nanotechnology
- Nanometrology and instrumentation
- Synthesis and processing

In Nature's words, Nature Nanotechnology "is a multidisciplinary journal that publishes papers of the highest quality and significance in all areas of nanoscience and nanotechnology. The journal covers research into the design, characterisation and production of structures, devices and systems that involve the manipulation and control of materials and phenomena at atomic, molecular and macromolecular scales. Both bottom-up and top-down approaches—and combinations of the two—are covered." The topics included are listed in Table 3.

Meanwhile, the USA Patent and Trademark Office, trying to reorganise to deal with nanotechnology patents and simplify nanotech investment, has it's own definition:

> "Nanotechnology is related to research and technology development at the atomic, molecular or macromolecular levels, in the length of scale of approximately 1-100 nanometer range in at least one dimension; that provide a fundamental understanding of phenomena and materials at the nanoscale; and to create and use structures, devices and systems that have novel properties and functions because of their small and/or intermediate size."

The FDA, concerned with human exposure, takes a similar approach.

> "The FDA defines "nanotechnology" as research and technology or development of products regulated by the FDA that involve all of the following:
>
> 1) the existence of materials or products at the atomic, molecular or macromolecular levels, where at least one dimension that affects the functional behaviour of the drug/device product is in the length scale range of approximately 1-100 nanometers;
>
> 2) the creation and use of structures, devices and systems that have novel properties and functions because of their small size; and,
>
> 3) the ability to control or manipulate the product on the atomic scale."

These definitions are so broad as to encompass virtually all developments in materials science and biotechnology. For example, although most of them make a point of requiring a "new" property to be considered, what is new? It's not clear whether simply making something small (even in only one dimension) is enough to be considered nanotechnology. Consider two additional contrarian definitions. Two of the most vocal non-profits critical of unfettered nanotechnology investment are Greenpeace and ETC Group. Greenpeace's definition is:

> "The most common definition of nanotechnology is that of manipulation, observation and measurement at a scale of less than

> 100 nanometers (one nanometer is one millionth of a millimeter). However, the emergence of a multi-disciplinary field called 'nanotechnology' arises from new instrumentation only recently available, and a flow of public money into a great number of techniques and relevant academic disciplines in what has been described as an 'arms race' between governments. Nanotechnology is really a convenient label for a variety of scientific disciplines which serves as a way of getting money from Government budgets."

The ETC Group definition:

> "Nanotechnology refers to the manipulation of living and non-living matter at the level of the nanometer (nm), one billionth of a meter. It is at this scale that quantum physics takes over from classical physics and the properties of elements change character in novel and unpredictable ways."

The ETC Group definition is focused only on the quantum nature of some nanoscale phenomena, accounting for only a part of the reality side of nanotechnology. The Greenpeace definition is very politically astute. It's interesting though that neither include in their definition anything about risk. It will prove interesting to check the longevity of these definitions a decade from now. They are evidence of the shotgun approach the professional community is taking toward nanotechnology. There has been some backlash to this from some of the scientific community. One example is a recent Nature commentary by French nanoscientist Christian Joachim. He suggests defining nano as:

> "Nanoscience should be reserved solely for the study of a single atom or a single molecule, that is, of one entity at a time, not for groups of such entities where statistics or interactions between them come into play."

This too narrow definition has good intentions. The ambiguity of the term nanotechnology sometimes attracts students with misperceptions about what they are getting into. Joachim suggests that a student might find themselves in a "physics of microelectronic devices" class instead of one on imagined nanomachines. Of course, one could argue that attracting students to take a physics class is good, misperception or not.

Perspectives from Science

Most nanotechnology definitions are cast in the language of application. This is good, since we are trying to describe a technology that does have applications now and in the future. Indeed, the language of application may be what unites all the subfields that make up nanotech and makes it a viable societal construct, though not necessarily a scientific one. Nevertheless, there are serious reasons not to lump too broad an array of advanced technologies under one banner. Education is one example.

Patenting is another. Labelling inventions "nanotechnologies" sounds like a good idea for promoting interest and investment in a new and promising arena, but divorcing a patent from the specific field or principle it derives from may be devastating to the patent process. Adequately judging prior work, quality, and worthiness may depend crucially on expertise in a specific field. There exist no experts in nanotechnology, nor will there, as society has defined the term.

Here we want to comment on the "new"-ness of "new nanoscale phenomena", which is central to so many of the definitions of nanotech above. Condensed matter (CM) physics, formerly known as solid-state physics, is the largest subfield of physics, the modern six being astrophysics, atomic physics, biophysics, condensed matter or solid state physics, high energy physics, and nuclear physics. CM physics deals with the study of matter that is condensed, including the physics of metals, semiconductors, liquids, and emergent phenomena in many particle systems.

CM seeks to understand the non-classical behaviour of everything, solid and liquid, that's cold from the scale of a few atoms to something you can hold in your hand. So its purview is quite broad and overlaps very much with materials science and physical chemistry. In the CM community there is very little change of behaviour due to the introduction of the word nanotechnology except indirectly via funding programmes. Here, emphasis on applied and interdisciplinary work has become more prevalent in recent years.

CM physics has ridden (and driven) the trend in the electronics industry towards miniaturisation to interesting discoveries at the nanoscale, including quantisation of electrical and thermal conductance and coulomb blockade in lateral quantum dots, though often at extremely low temperatures. When CM physicists consider nanotechnology or nanoscience they usually think of one of three things:

1) quantum size effects that emerge at the nanoscale,
2) new surface physics/energetics/mechanics in nanoscale structures due to the increased surface to volume ratio, and
3) quantum coherent transport effects through mesoscale systems.

Largely these occur in metallic and crystallographic systems, the most common of the latter being silicon, gallium arsenide, and carbon. There is also a history of studying extremely small life "nanno-biology" but this history has been largely forgotten. The nano-bio-technology we see today in electrical engineering labs that are going "wet" is really a merger of hard inorganic device (CM) physics being applied to biological matter.

Cold or Hot, quantum or not

Probably the most economically important and well-studied material on earth, it has taken almost 100 years to almost fully understand this simple periodic structure. Consider an atom of silicon. It has discrete energy levels or "shells" where electrons can exist. As the atoms come together to form a crystal, the energy levels become energy bands, continuous as a function of the electron's position within the crystal lattice.

Since silicon is a semiconductor, its optical properties are largely governed by a band-gap that separates the valence electron bands from the conduction bands, where the electrons are much more mobile. Semiconductor technology is quantum, in some sense, as it requires a quantum mechanical description of matter to explain. The progression of a block of silicon crystal from the macroscopic scale down is a great way to understand some of the key manifestations of new nanoscale phenomena, but we could use many systems.

Nanotechnology proponents often bring up new quantum behaviour that emerges at the nanoscale. There is nothing new, per se, about this phenomenon—being governed by well-known quantum physics—but only recently could matter be controlled well enough to reach this scale. There are two ways in general to see quantum effects:

make a material colder or make it smaller. This is because the properties of materials, from atoms to molecules to metals and crystals, are usually governed by the interaction of electrons. And the properties of electrons are governed by the energy levels where they are allowed to exist.

Crudely speaking, temperature blurs these energy levels so that electrons can hop from one level to another more easily. Here is the key point: if we make a material small enough, these energy levels will move farther apart, making this hopping due to temperature much more difficult. So, in some sense, making an object very small makes it "colder." Other new properties appear as well, such as a change in the optical excitation response—as this is defined by the spacing of these energy levels as well. This phenomena, which happens to take place around the nano-scale (say 10 nm for silicon) is what is often referred to as "quantum size effects."

Nano in Nanoparticles

Consider nanoparticles, the nanotechnology of the present. These include nanocrystals made of either metal or semiconductor as well as the carbon fullerene family, including buckyballs and nanotubes. Nanowires are classified under nanocrystals since they are solid objects as opposed to hollow nanotubes. All nanoparticles have at least two dimensions in the nanoscale regime, 1-100 nm. They have new properties that appear because of their size that would disappear in the micron regime. These "new properties" can be different in origin and even coincidental. Categorise them broadly into three bins:

1) new quantum effects which appear due to confinement, which are often called quantum size effects;
2) new surface physics or much increased reactivity, usually attributed to increased surface to volume ratios—this might also include structural changes from adding nanoparticles to a material (like rubber) to make it stronger; and
3) new biological properties which are largely a coincidence of scale, as the body is not attuned to defend against nanosised particles in many situations.

It would be easy to argue that nanoparticles or quantum dots are the only nanotechnology at this point in time. They certainly constitute most of the risk. Nanoparticles have large surface-to-volume ratios making them extremely reactive. They also are able to penetrate sensitive areas of the body and accumulate there.

They can be embedded in larger substrates such as polymer systems or fabrics, which may decay over time and release these nanoparticles into the environment. They are also found in sunscreens and cosmetics. Nanoparticles and (eventually) active nanomachines bridge the materials to biology gap. Evolution in biomarkers for medical imaging and cancer treatment has tracked well with nanoparticle technology. But not all quantum dots are dangerous.

Not Nanotechnology

Technologies are often organised by scientific principle or concept, whereas nanotech can incorporate these just because of a coincidence in scales. Take the case of photonic band gap materials, which apply the theories historically developed for electronic waves in semiconductors to photon waves (light) travelling in periodic index of refraction materials—sometimes called meta-materials. Visible light photonic band gap materials will likely require nanoscale manufactured devices (<100 nm 3D structures) but work as well at microwave frequencies in order 1 cm sized structures.

The physics doesn't change whether the features are 1 nm or 1 cm. So the only thing "nano" about some photonic meta-materials may be the fabrication process. Is there any point calling it nanotechnology? Would you send this patent application to be reviewed by the "nanotechnology" patent clerk or the physical optics patent clerk? It should be a no-brainer (the latter).

Devices or structures with only one dimension at the nanometer length scale are in general not nanotechnology. Quantum wells are a good example. To make a quantum well you sandwich a thin semiconductor layer between two bigger semiconductor layers of a different type. This allows you to trap electrons in the inner layer. Physically this is not much different from inversion layers, where one material is grown on the other and an electric field traps electrons at the interface. These kinds of semiconductor devices are well known.

Electrons in these 2D sheets exhibit many new and interesting phenomena, often attributed to the nanoscale. But, new nano phenomena in this context could just as easily be called new quantum phenomena or new manybody phenomena.

Superconductivity, quantum wells and electron gasses, giant magneto-resistance (GMR) are all examples of quantum behaviour misappropriated to the nano-field. This also explains some of the ridiculous money expenditure estimates—in the tens of billions of dollars some of them -which have been touted as nano-investment. If you include all the investment in GMR hard drive technology, you get big numbers.

References

Amato, I., "Nanotechnology: Shaping the world atom by atom", *National Science and Technology Council*, USA., 2000.

Colvin, V.L., "The Potential Environmental Impact of Engineered Nanomaterials", *Nature Biotechnology* 21(10): 1166-1170, 2003.

Feynman, R. P., *There's plenty of room at the bottom: An invitation toenter a new field of physics*, APS Meeting, Dec. 29, 1959.

Ghalanbor Z, Marashi SA, Ranjbar B., "Nanotechnology helps medicine: nanoscale swimmers and their future applications", *Med Hypotheses* 65 (1): 198-199, 2005.

Hatzor-de Picciotto, A., A. D. Wissner-Gross, G. Lavallee, P. S. Weiss, "Arrays of Cu(2+)-complexed organic clusters grown on gold nano dots". *Journal of Experimental Nanoscience* 2: 3-11, 2007.

Hood, E., *Nanotechnology: Looking As We Leap*, Environmental Health Perspectives, v112, 2004.

Joy, B., *Why the Future Doesn't Need Us*, *Wired* April, 8(4), 2000.

Masciangioli, T. and Zhang, W., *Environmental Technologies at the Nanoscale*, *Env Sci & Tech*, 102A-108A, March 1, 2003.

Stephenson, N., *The Diamond Age or, A Young Lady's Illustrated Primer*, Bantam Books, New York, 1995.

Sundrani D, Darling SB, Sibener SJ, "Hierarchical assembly and compliance of aligned nanoscale polymer cylinders in confinement", *Langmuir 20* (12): 5091-9, June 2004.

Tahan, C., *Science and Technology Studies 201: Nanotechnology and Society*, URL, 2006, http://www.tahan.com/charlie/nanosociety/course201/.

Ellison, W. C. Crone, and Clark A. Miller, "Nanotechnology and Society: A discussion-based undergraduate course", *Am. J. Phys.* 74, 4, April 2006.

Tahan, C., "The Nanotechnology R(evolution*)*", In Fritz Allhoff, Patrick Lin, James Moor, and John Weckert, editors, *Nanoethics: Examining the Societal Impact of Nanotechnology*, John Wiley & Sons, 2007.

Warheit, D., "Nanoparticles: Health Impacts?", *Materials Today* Feb. 2004.

9

Nanoelectronic Devices

Nanoelectronics are sometimes considered as disruptive technology because present candidates are significantly different from traditional transistors. Some of these candidates include: hybrid molecular/ semiconductor electronics, one dimensional nanotubes/nanowires, or advanced molecular electronics.

The sub-voltage and deep-sub-voltage nanoelectronics are specific and important fields of R&D, and the appearance of new ICs operating almost near theoretical limit (fundamental, technological, design methodological, architectural, algorithmic) on energy consumption per 1 bit processing is inevitable. The important case of fundamental ultimate limit for logic operation is reversible computing. Although all of these hold immense promises for the future, they are still under development and will most likely not be used for manufacturing any time soon.

APPROACHES TO NANOELECTRONICS

Use of Nanofabrication

For example, single electron transistors, which involve transistor operation based on a single electron. Nanoelectromechanical systems also falls under this category. Nanofabrication can be used to construct ultradense parallel arrays of nanowires, as an alternative to synthesizing nanowires individually.

Nanomaterials Electronics

Besides being small and allowing more transistors to be packed into a single chip, the uniform and symmetrical structure of nanotubes allows a higher electron mobility (faster electron movement in the material), a higher dielectric constant (faster frequency), and a symmetrical electron/hole characteristic. Also, nanoparticles can be used as quantum dots.

Concept of Molecular electronics

Molecular electronics (sometimes called moletronics) is an interdisciplinary theme that spans physics, chemistry, and materials science. The unifying feature of this area is the use of molecular building blocks for the fabrication of electronic components, both passive and active. The concept of molecular electronics has aroused much excitement both in science fiction and among scientists due to the prospect of size reduction in electronics offered by molecular-level control of properties.

Molecular electronics provides means to extend Moore's Law beyond the foreseen limits of small-scale conventional silicon integrated circuits. Due to the broad use of the term, molecular electronics can be split into two related but separate subdisciplines: molecular materials for electronics utilises the properties of the molecules to affect the bulk properties of a material, while molecular scale electronics focuses on single-molecule applications.

Concept Genesis and Theory

Study of charge transfer in molecules was advanced in the 1940s by Robert Mulliken and Albert Szent-Gyorgi in discussion of so-called "donor-acceptor" systems and developed the study of charge transfer and energy transfer in molecules. Likewise, a 1974 paper from Mark Ratner and Ari Aviram 1 illustrated a theoretical molecular rectifier. Later, in 1988, Aviram described in detail a theoretical single-molecule field-effect transistor.

Further concepts were proposed by Forrest Carter of the Naval Research Laboratory, including single-molecule logic gates.

Unfortunately, the direct measurement of the electronic characteristics of individual molecules awaited the development of methods for making molecular-scale electrical contacts. This was no easy task.

Thus, the first experiment measuring the conductance of a single molecule was only reported in 1997 by Mark Reed and co-workers. Since then, this branch of the field has progressed rapidly. Likewise, as it has become possible to measure such properties directly, the theoretical predictions of the early workers have been mostly confirmed.

Figure 1. Voltage-controlled switch, a molecular electronic device from 1974.

However, while mostly operating in the quantum realm of less than 100 nanometers, "molecular" electronic processes often collectively manifest on a macro scale. Examples include quantum tunnelling, negative resistance, phonon-assisted hopping, polarons, and the like. Thus, macro-scale active devices were described decades before molecular-scale ones. E.g., in 1974, John McGinness and his coworkers described the putative "first experimental demonstration of an operating molecular electronic device". This was a voltage-controlled switch.

As its active element, this device used DOPA melanin, an oxidised mixed polymer of polyacetylene, polypyrrole, and polyaniline. The "ON" state of this switch exhibited almost metallic conductivity. Since the 1970's, scientists have developed an entire panoply of new materials and devices. These findings have opened the door to plastic electronics and optoelectronics, which are beginning to find commercial application.

Charge Transfer Complexes

The first highly-conductive organic compounds were the Charge transfer complexes. In 1954, researchers at Bell Labs and elsewhere reported Charge transfer complexes with resistivities as low as 8 ohms-cm. In the early 1970's, salts of tetrathiafulvalene were shown to exhibit almost metallic conductivity, while superconductivity was demonstrated in 1980. Broad research on charge transfer salts continues today.

Conducting Polymers

The linear-backbone "polymer blacks" (polyacetylene, polypyrrole, and polyaniline) and their copolymers are the main class of conductive polymers. Historically, these are known as Melanins. In 1963 Australians DE Weiss and coworkers reported iodine-doped oxidised polypyrrole blacks with resistivities as low as 1 ohm/cm.

Subsequent papers reported resistances as low as 0.03 Ohm/cm. With the notable exception of Charge transfer complexes, organic molecules had previously been considered insulators or at best weakly conducting semiconductors.

Over a decade later in 1977, Shirakawa, Heeger, and MacDiarmid reported equivalent high conductivity in rather similarly oxidised and iodine-doped polyacetylene. They later received the 2000 Nobel prize in chemistry for "The discovery and development of conductive polymers ". The Nobel citation made no reference to Weiss et al's similar earlier work. Also see Nobel Prize controversies.

C_{60} AND CARBON NANOTUBES

Graphite to C_{60}

In polymers, classical organic molecules are composed of both carbon and hydrogen (and sometimes additional compounds such as nitrogen, chlorine or sulphur). They are obtained from petrol and can often be synthethized in large amounts. Most of these molecules are insulating when their length exceeds a few nanometers. However, naturally occurring carbon is conducting.

In particular, graphite (recovered from coal or encountered naturally) is conducting. From a theoretical point of view, graphite is a semi-metal, a category in between metals and semi-conductors. It has a layered structure, each sheet being one atom thick. Between each sheet, the interactions are weak enough to allow an easy manual cleavage.

Tailoring the graphite sheet to obtain well defined nanometer-sized objects remains a challenge. However, by the close of the twentieth century, chemists were exploring methods to fabricate extremely small graphitic objects that could be considered single molecules. After studying the interstellar conditions under which carbon is known to form clusters, Richard Smalley's group set up an experiment in which graphite was vaporised using laser irradiation.

Mass spectrometry revealed that clusters containing specific "magic numbers" of atoms were stable, in particular those clusters of 60 atoms. Harry Kroto, an English chemist who assisted in the experiment, suggested a possible geometry for these clusters—atoms covalently bound with the exact symmetry of a soccer ball.

Coined buckminsterfullerenes, buckyballs or C_{60}, the clusters retained some properties of graphite, such as conductivity. These objects were rapidly envisioned as possible building blocks for molecular electronics.

THEORY OF MOLECULAR ELECTRONICS

The theory of single molecule devices is particularly interesting since the system under consideration is an open quantum system in nonequilibrium (driven by voltage).

In the low bias voltage regime, the nonequilibrium nature of the molecular junction can be ignored, and the current-voltage characteristics of the device can be calculated using the equilibrium electronic structure of the system. However, in stronger bias regimes a more sophisticated treatment is required, as there is no longer a variational principle.

In the elastic tunnelling case, the formalism of Rolf Landauer can be used to calculate the transmission through the system as a function of bias voltage, and hence the current.

In inelastic tunnelling, an elegant formalism based on the non-equilibrium Green's functions of Leo Kadanoff and Gordon Baym, and independently by Leonid Keldysh was put forth by Ned Wingreen and Yigal Meir. This Meir-Wingreen formulation has been used to great success in the molecular electronics community to examine the more difficult and interesting cases where the transient electron exchanges energy with the molecular system.

Recent progress in nanotechnology and nanoscience has facilitated both experimental and theoretical study of molecular electronics. In particular, the development of the scanning tunnelling microscope (STM) and later the atomic force microscope (AFM) have facilitated manipulation of single-molecule electronics.

The first widely reported result was using a mechanical break junction approach to connect two gold electrodes to a sulphur-terminated molecular wire by Mark Reed and James Tour.

A collaboration of researchers at HP and UCLA, led by James Heath, Fraser Stoddart, R. Stanley Williams, and Philip Kuekes, has developed molecular electronics based on rotaxanes and catenanes. Work is also being done on the use of single-wall carbon nanotubes as field-effect transistors. Most of this work is being done by IBM.

The Aviram-Ratner model for a molecular rectifier, which until recently was entirely theoretical, has been confirmed experimentally and unambiguously i a number of experiments by a group led by Geoffrey J. Ashwell at Cranfield University, UK. Many rectifying molecules have so far been identified, and the number and efficiency of these systems is expanding rapidly.

Supramolecular electronics is a new field that tackles electronics at a supramolecular level. An important issue in molecular electronics is the determination of the resistance of a single molecule (both theoretical and experimental). For example, Bumm, et al used STM to analyse a single molecular switch in a self-assembled monolayer to determine how conductive such a molecule can be. Another problem faced by this field is the difficulty to perform direct characterisation since imaging at the molecular scale is often difficult in many experimental devices.

DEFINE SINGLE MOLECULE ELECTRONICS

Single molecule electronics is a branch of molecular electronics that uses single molecules as electronic components. Because single molecules constitute the smallest stable structures imaginable this miniaturisation is the ultimate goal for shrinking electrical circuits.

Concepts of Single Molecule Electronics

In single molecule electronics, the bulk material is replaced by single molecules. That is, instead of creating structures by removing or applying material after a pattern scaffold, the atoms are put together in a chemistry lab. This way billions of billions of copies are made simultaneously (typically more than 10 20 molecules are made at once) while the composition of molecules are controlled down to the last atom. The molecules utilised have properties that resemble traditional electronic components such as a wire, transistor or rectifier.

The miniaturisation down to single molecules brings the scale down to a regime where quantum effects are important. As opposed to the case in conventional electronic components, where electrons can be filled in or drawn out more or less like a continuous flow of charge, the transfer of a single electron alters the system significantly. This means that when an electron has been transferred from the source electrode to the molecule, the molecule gets charged up and makes it much harder for the next one to transfer.

The significant amount of energy due to charging has to be taken into account when making calculations about the electronic properties of the setup and is highly sensitive to distances to conducting surfaces nearby.

Because of the small size of the molecules, quantum mechanics put severe restrains on the states (or orbitals) the electrons can be in on the molecule. These states determine the energy and spatial distribution that an electron can have and hence the electronic properties of the setup.

Unfortunately, even though the molecules seem small and simple when drawn schematically, the possible electronic states can only be deduced approximately and this limits the predictability of the molecular electronic properties. Further, connecting single molecules reliably to a larger scale circuit has proven to be a great challenge and constitute a significant hindrance to commercialisation.

Conventionally, the electronics are made up of bulk material. Ever since its invention in 1958 the performance and complexity of integrated circuits has been growing exponentially and has forced the feature sizes of the embedded components to shrink accordingly.

As the structures become smaller the sensitivity for deviations increases and in a few generations, when the minimum feature sizes reaches 13 nm, the composition of the devices will have to be controlled to a precision of a few atoms in order for the devices to work.

With the bulk approach having inherent limitations in addition to becoming increasingly demanding and expensive, the idea was born that the components could instead be build up atom for atom in a chemistry lab as opposed to carving them out of bulk material. This idea is the reasoning behind molecular electronics with the ultimate miniaturisation being components contained in single molecules.

The concept of molecular electronics was first published in 1974 when Aviram and Ratner suggested an organic molecule that could work as a rectifier. Having both huge commercial and fundamental interest much effort was put into proving its feasibility and 16 years later in 1990 the first demonstration of an intrinsic molecular rectifier was realised by Ashwell and coworkers for a thin film of molecules. It wasn't until 1997 that the first measurements on the conductance of a single molecule were published.

Examples of Molecular Electronics

Common for molecules utilised in molecular electronics is that the structures contain a lot of alternating double and single bonds. The reason for this is that such a pattern delocalises the molecular orbitals making it possible for electrons to move freely over the conjugated area.

Molecular wires

The sole purpose of molecular wires is to electrically connect different parts of a molecular electrical circuit.

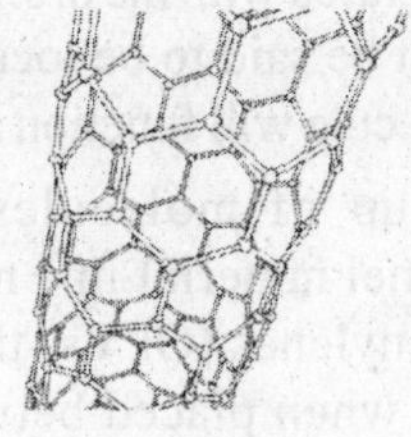

Figure 2. This animation of a rotating carbon nanotube shows its 3D structure.

As the assembly of these and their connection to a macroscopic circuit is still to be mastered, the focus of research in single molecule electronics is primarily on the functionalised molecules: molecular wires are characterised by containing no functional groups and are hence composed of plain repetitions of a conjugated building block.

Among these are the carbon nanotubes that are quite large compared to the other suggestions but have shown very promising electrical properties. The main problem with the molecular wires is to obtain good electrical contact with the electrodes so that the electrons can move freely in and out of the wire.

Molecule transistors

Single molecule transistors are fundamentally different than the ones known from bulk electronics. The gate in a conventional (field-emission) transistor determines the conductance between the source and drain electrode by controlling the density of charge carriers between them, whereas the gate in a single molecule transistor controls the feasibility of a single electron to jump on and off the molecule by modifying the energy of the molecular orbitals.

One of the effects of this difference is that the single molecule transistor is almost binary: it is either ON or OFF. This opposes its bulk counterparts which have quadratic responses to gate voltage. It is the quantisation of charge into electrons that is responsible for the markedly different behaviour compared to bulk electronics. Because of the size of a single molecule, the charging due to a single electron is significant and provides a mean to turn the transistor ON or OFF.

For this to work, the electronic orbitals on the transistor molecule cannot be too well integrated with the orbitals on the electrodes. If they are, an electron cannot be said to be located on the molecule or the electrodes and the molecule will function as a wire.

A popular group of molecules, that can work as the semiconducting channel material in a molecular transistor, is the oligopolyphenylenevinylenes (OPVs) that works by the Coulomb blockade mechanism when placed between the source and drain electrode in an appropriate way. Fullerenes work by the same mechanism and have also been commonly utilised.

Semiconducting carbon nanotubes have also been demonstrated to work as channel material but although molecular, these molecules are sufficiently large to behave almost as bulk semiconductors.

The size of the molecules and the low temperature the measurements are being conducted at makes the quantum mechanical states well defined. It is therefore being researched if the quantum mechanical properties can be used for more advanced purposes than simple transistors.

Physicists at the University of Arizona, in collaboration with chemists from the University of Madrid, have designed a single molecule transistor using a ring-shaped molecule similar to benzene. Physicists at Canada's National Institute for Nanotechnology have designed a single-molecule transistor using styrene. Both groups expect their respective devices to function at room temperature, and to be controlled by a single electron.

Rectifiers

Molecular rectifiers are mimics of their bulk counterparts and have an asymmetric construction so that the molecule can accept electrons in one end but not the other. The molecules have an electron donor (D) in one end and an electron acceptor (A) in the other. This way, the unstable state D^+ - A^- will be more readily made than D^- - A^+. The result is that an electric current can be drawn through the molecule if the electrons are added through the acceptor end, but not so easily if the reverse is attempted.

Techniques

One of the biggest problems with measuring on single molecules is to establish reproducible electrical contact with only one molecule and doing so without shortcutting the electrodes. Because the current photolithographic technology is unable to produce electrode gaps small enough to contact both ends of the molecules tested (in the order of nanometers) alternative strategies is put into use.

Molecular gaps

One way to produce electrodes with a molecular sized gap between them is break junctions, in which a thin electrode is stretched until it breaks. Another is electromigration. Here a current is lead through a thin wire until it melts and the atoms migrate to produce the gap. Further, the reach of conventional photolithography can be enhanced by chemically etching or depositing metal on the electrodes.

Probably the easiest way to conduct measurements on several molecules is to use the tip of a scanning tunnelling microscope (STM) to contact molecules adhered at the other end to a metal substrate. The connection obtained this way is, however, not stable enough to conduct the advanced measurements possible with the above techniques.

Anchoring

A popular way to anchor molecules to the electrodes is to make use of sulfurs' high affinity to gold. In these setups, the molecules are synthesized so that sulphur atoms are placed strategically to function as crocodile clips connecting the molecules to the gold electrodes. Though useful, the anchoring is non-specific and thus anchors the molecules randomly to all gold surfaces.

Further, the contact resistance is highly dependent on the precise atomic geometry around the site of anchoring and thereby inherently compromises the reproducibility of the connection. To circumvent the latter issue, experiments has shown that fullerenes could be a good candidate for use instead of sulphur because of the large conjugated p-system that can electrically contact many more atoms at once than a single atom of sulphur.

Problems of Molecules

Artifacts

When trying to measure electronic characteristics of molecules, artificial phenomena can occur that can be hard to distinguish from truly molecular behaviour. Before they were discovered these artifacts have mistakenly been published as being features pertaining to the molecules in question. Applying a voltage drop in the order of volts across a nanometer sized junction results in a very strong electrical field.

The field can cause metal atoms to migrate and eventually close the gap by a thin filament which can be broken again when carrying a current. The two levels of conductance imitate molecular switching between a conductive and an isolating state of a molecule.

Another encountered artifact is when the electrodes undergo chemical reactions due to the high field strength in the gap. When the bias is reversed the reaction will cause hysteresis in the measurements that can be interpreted as being of molecular origin.

A metallic grain between the electrodes can act as a single electron transistor by the mechanism described above thus resembling the characteristics of a molecular transistor. This artifact is especially common with nanogaps produced by the electromigration technique.

Commercialisation

One of the biggest hindrances for single molecule electronics to be commercially exploited is the lack of techniques to connect a molecular sized circuit to bulk electrodes in a way that gives reproducible results. At the current state, the difficulty of connecting single molecules vastly outweighs any possible performance increase that could be gained from such shrinkage. The picture becomes even worse if the molecules are to have a certain spatial orientation and/or have multiple poles to connect.

Also problematic is the fact that measurements on single molecules are carried out in cryogenic temperatures (close to absolute zero) which is very energy consuming. This is done in order to reduce the signal noise to a degree where the faint currents of single molecules can be measured.

Perspectives

Single molecule electronics is an emerging field, and entire electronic circuits consisting exclusively of molecular sized compounds are still very far from being realised. However, the continuous demand for more computing power together with the inherent limitations of the present day lithographic methods make the transition seem unavoidable. Currently, the focus is on discovering molecules with interesting properties and on finding ways to obtaining reliable and reproducible contacts between the molecular components and the bulk material of the electrodes.

Supramolecular Electronics

Supramolecular electronics is the experimental field of supramolecular chemistry that bridges the gap between molecular electronics and bulk plastics in the construction of electronic circuitry at the nanoscale. In supramolecular electronics, assemblies of pi-conjugated systems on the 5 to 100 nanometer length scale are prepared by molecular self-assembly with the aim to fit these structures between electrodes.

With single-molecules as researched in molecular electronics at the 5 nanometer scale this would be impractical. Nanofibers can be prepared from polymers such as polyaniline and polyacetylene. Chiral oligo (p-phenylenevinylene)s self-assemble in a controlled fashion into (helical) wires.

NANOCIRCUITS

Nanocircuits are electrical circuits on the scale of nanometers. One nanometer is equal to 10^{-9} meters or a row of 10 hydrogen atoms. With circuits becoming smaller, there is an ability to fit more on a computer chip. This allows more complex functions using less power and at a faster speed. Nanocircuits are organised into three different parts: transistors, interconnections, and architecture, all dealt with on the nano scale.

A variety of proposals have been made to implement nanocircuitry in different forms including Single-Electron Transistors, Quantum dot cellular automata, and Nanoscale Crossbar Latches. However, it is likely that near term approaches will be based on the incorporation of nanomaterials to improve the design of MOSFETs, which currently form the basis of most analog and digital circuit designs and the scaling of which drives Moore's Law.

The MOSFET design and its future was published in 2004 comparing different geometries of MOSFETs under scale reduction and noted that circular cross-section vertical channel FETs are optimal for scale reduction. This configuration is capable of being implemented with a high density using vertical semiconductor cylindrical channels with nanoscale diameters and Infineon Technologies and Samsung have begun research and development in this direction resulting in some basic patents using nanowires and carbon nanotubes in MOSFET designs.

In an alternative approach, Nanosys is a new company using solution based deposition and alignment processes to pattern pre-fabricated arrays of nanowires on a substrate to serve as a lateral channel of an FET. While not capable of the same scalability as single nanowire FETs, the use of pre-fabricated multiple nanowires for the channel increases reliability and reduces production costs since large volume printing processes may be used to deposit the nanowires at a lower temperature than conventional fabrication procedures.

In addition, due to the lower temperature deposition a wider variety of materials such as polymers may be used as the carrier substrate for the transistors opening the door to flexible electonics applications such as electronic paper, bendable flat panel displays, and wide area solar cells.

Production Methods

One of the most fundamental concepts to understanding nanocircuits is the formulation of Moore's Law. This concept arose when Intel co-founder Gordon Moore became interested in the cost of transistors and trying to fit more onto one chip. It relates that the number of transistors that can be fabricated on a silicon integrated circuit-and therefore the computing speed of such a circuit-is doubling every 18 to 24 months.

The more transistors one can fit on a circuit, the faster the computer will be. This is why scientists and engineers are working together to produce these nanocircuits so millions and perhaps even billions of transistors will be able to fit onto a chip. Despite how good this may sound, there are many problems that arise when so many transistors are packed together. With circuits being so tiny, they tend to have more problems than larger circuits, more particularly many defects.

Nanoscale circuits are more sensitive to temperature changes, cosmic rays and electromagnetic interference than today's circuits. As they pack more transistors onto a chip, phenomena such as stray signals on the chip, the need to dissipate the heat from so many closely packed devices, and the difficulty of creating the devices in the first place will halt or severely slow progress.

Many believe the market for nanocircuits will reach equilibrium around 2015. At this time they believe the cost of a fabrication facility may be as much as $200 billion. There will be a time when the cost of making circuits even smaller will be too much, and the speed of computers will reach a maximum.

For this reason, many scientists believe that Moore's Law will not hold forever and will soon reach a peak. In producing these nanocircuits, there are many aspects involved. The first part of their organisation begins with transistors. As of right now, most electronics are using silicon-based transistors.

Transistors are an integral part of circuits as they control the flow of electricity and transform weak electrical signals to strong ones. They also control electric current as they can turn it on off, or even amplify signals. Circuits now use silicon as a transistor because it can easily be switched between conducting and nonconducting states. However, in nanoelectronics, transistors might be organic molecules or nanoscale inorganic structures.

Semiconductors, which are part of transistors, are also being made of organic molecules in the nano state. The second aspect of nanocircuit organisation is interconnection. This involves logical and mathematical operations and the wires linking the transistors together that make this possible.

In nanocircuits, nanotubes and other wires as narrow as one nanometer are used to link transistors together. Nanowires have been made from carbon nanotubes for a few years. Until a few years ago, transistors and nanowires were put together to produce the circuit. However, scientists have been able to produce a nanowire with transistors in it.

In 2004, Harvard University nanotech pioneer Charles Lieber and his team have made a nanowire-10,000 times thinner than a sheet of paper-that contains a string of transistors. Essentially, transistors and nanowires are already pre-wired so as to eliminate the difficult task of trying to connect transistors together with nanowires.

The last part of nanocircuit organisation is architecture. This has been explained as the overall way the transistors are interconnected, so that the circuit can plug into a computer or other system and operate independently of the lower-level details. With nanocircuits being so small, they are destined for error and defects.

Scientists have devised a way to get around this. Their architecture combines circuits that have redundant logic gates and interconnections with the ability to reconfigure structures at several levels on a chip. The redundancy lets the circuit identify problems and reconfigure itself so the circuit can avoid more problems. It also allows for errors within the logic gate and still have it work properly without giving a wrong result.

Potential Applications

Scientists in India have recently developed the world's smallest transistor which will be used for nanocircuits. The transistor is made entirely from carbon nanotubes. Nanotubes are rolled up sheets of carbon atoms and are more than a thousand times thinner than human hair. Normally circuits use silicon-based transistors, but these will soon replace those. The transistor has two different branches that meet at a single point, hence giving it a Y shape.

Current can flow throughout both branches and is controlled by a third branch that turns the voltage on or off. This new breakthrough can now allow for nanocircuits to hold completely to their name as they can be made entirely from nanotubes. Before this discovery, logic circuits used nanotubes, but needed metal gates to be able to control the flow of electrical current.

Arguably the biggest potential application of nanocircuits deals with computers and electronics. Scientists and engineers are always looking to make computers faster. Some think in the nearer term, we could see hybrids of micro and nano: silicon with a nano core-perhaps a high-density computer memory that retains its contents forever.

Unlike conventional circuit design, which proceeds from blueprint to photographic pattern to chip, nanocircuit design will probably begin with the chip-a haphazard jumble of as many as 1024 components and wires, not all of which will even work-and gradually sculpt it into a useful device.

Instead of taking the traditional top-down approach, the bottom-up approach will probably soon have to be adopted because of the sheer size of these nanocircuits. Not everything in the circuit will probably work because at the nano level, nanocircuits will be more defective and faulty because of their compactness. Scientists and engineers have created all of the essential components of nanocircuits such as transistors, logic gates and diodes. They have all been constructed from organic molecules, carbon nanotubes and nanowire semiconductors.

The only thing left to do is find a way to eliminate the errors that come with such a small device and nanocircuits will become a way of all electronics. However, eventually there will be a limit as to how small nanocircuits can become and computers and electronics will reach their equilibrium speeds.

Economic Impact

With the vast improvements in reducing the size of circuits, comes a rising cost to produce these nano components. Scientists believe that one day a fabrication facility for making nanocircuit could cost as much as over $200 billion. The increased cost comes from the difficulty of producing such circuits as they take more time and effort than circuits today. The fabrication plant will create a raw nanocircuit-billions on billions of devices and wires whose functioning is rather limited. From the outside it will look like a lump of material with a handful of wires sticking out.

Eventually the theory of Moore's Law will have to reach equilibrium with the fabrication methods currently used. Circuits will only be able to be so fast and small without creating any severe problems. The cost for producing even better nanocircuits will increase further as more money will be needed to develop new fabrication methods and ways of designing faster, better nanocircuits.

Until that time, companies like Intel will continue to thrive in the nano business with their promises of their chip being the fastest and better than their counterpart. Nanocircuits may still have their problems, but that will not stop companies from mass producing them in order to become the most technologically advanced company with the fastest product.

Nanoelectronic Devices

Radios

Nanoradios have been developed structured around carbon nanotubes. A nanoradio is a radio receiver or transmitter constructed on a nanometer scale. Currently only receivers have been developed and they are structured around a carbon nanotube. The first such device was described in October 2007 by a team led by physicist Alex Zettl.

The nanotube, about 10 nanometers in diameter and several hundred nanometers long, is contained in a vacuum and one of its ends is connected to an electrode of a battery. The other electrode is placed a short distance from the nanotube's other end. The tube, now charged, will vibrate in tune with any external electromagnetic signal, effectively acting as an antenna.

The favourable vibration frequency can be adjusted by changing the applied voltage, which allows to tune the radio to different carrier frequencies. The field emission effect causes a current to flow, as electrons tunnel across the gap between the tube and the second electrode. This current represents an amplified version of the radio signal; no demodulation is necessary.

Computers

Nanoelectronics holds the promise of making computer processors more powerful than are possible with conventional semiconductor fabrication techniques. A number of approaches are currently being researched, including new forms of nanolithography, as well as the use of nanomaterials such as nanowires or small molecules in place of traditional CMOS components. Field effect transistors have been made using both semiconducting carbon nanotubes and with heterostructured semiconductor nanowires.

Energy production

Research is ongoing to use nanowires and other nanostructured materials with the hope of to create cheaper and more efficient solar cells than are possible with conventional planar silicon solar cells. It is believed that the invention of more efficient solar energy would have a great effect on satisfying global energy needs.

There is also research into energy production for devices that would operate in vivo, called bio-nano generators. A bio-nano generator is a nanoscale electrochemical device, like a fuel cell or galvanic cell, but drawing power from blood glucose in a living body, much the same as how the body generates energy from food. To achieve the effect, an enzyme is used that is capable of stripping glucose of its electrons, freeing them for use in electrical devices.

The average person's body could, theoretically, generate 100 watts of electricity using a bio-nano generator. However, this estimate is only true if all food was converted to electricity, and the human body needs some energy consistently, so possible power generated is likely much lower. The electricity generated by such a device could power devices embedded in the body (such as pacemakers), or sugar-fed nanorobots.

A similar technology was presented in the Matrix series of science fiction major motion pictures, with robots shown enslaving mankind for its bio-energy. Much of the research done on bio-nano generators is still experimental, with Panasonic's Nanotechnology Research Laboratory among those at the forefront.

Medical diagnostics

There is great interest in constructing nanoelectronic devices that could detect the concentrations of biomolecules in real time for use as medical diagnostics, thus falling into the category of nanomedicine. A parallel line of research seeks to create nanoelectronic devices which could interact with single cells for use in basic biological research. These devices are called nanosensors. Such miniaturisation on nanoelectronics towards in vivo proteomic sensing should enable new approaches for health monitoring, surveillance, and defence technology.

Nanowires

A nanowire is a wire of diameter of the order of a nanometer (10^{-9} meters). Alternatively, nanowires can be defined as structures that have a lateral size constrained to tens of nanometers or less and an unconstrained longitudinal size. At these scales, quantum mechanical effects are important—hence such wires are also known as "quantum wires".

Many different types of nanowires exist, including metallic, semiconducting, and insulating. Molecular nanowires are composed of repeating molecular units either organic or inorganic. The nanowires could be used, in the near future, to link tiny components into extremely small circuits. Using nanotechnology, such components could be created out of chemical compounds.

Typical nanowires exhibit aspect ratios (length-to-width ratio) of 1000 or more. As such they are often referred to as 1-Dimensional materials. Nanowires have many interesting properties that are not seen in bulk or 3-D materials. This is because electrons in nanowires are quantum confined laterally and thus occupy energy levels that are different from the traditional continuum of energy levels or bands found in bulk materials.

Peculiar features of this quantum confinement exhibited by certain nanowires such as carbon nanotubes manifest themselves in discrete values of the electrical conductance. Such discrete values arise from a quantum mechanical restraint on the number of electrons that can travel through the wire at the nanometer scale. These discrete values are often referred to as the quantum of conductance and are integer values of

$$\approx 12.9\ k\Omega^{-1}$$

They are inverse of the well-known resistance unit h/e^2, which is roughly equal to 25812.8 ohms, and referred to as the von Klitzing constant R_K (after Klaus von Klitzing, the discoverer of exact quantization). Since 1990, a fixed conventional value R_{K-90} is accepted. Examples of nanowires include inorganic molecular nanowires ($Mo_6S_{9-x}I_x$, $Li_2Mo_6Se_6$), which have a diameter of 0.9 nm, and can be hundreds of micrometers long. Other important examples are based on semiconductors such as InP, Si, GaN, etc., dielectrics, or metals.

There are many applications where nanowires may become important in electronic, opto-electronic and nanoelectromechanical devices, as additives in advanced composites, for metallic interconnects in nanoscale quantum devices, as field-emitters and as leads for biomolecular nanosensors.

Physics of nanowires

Nanowires are not observed spontaneously in nature and must be produced in a laboratory. Nanowires can be either suspended, deposited or synthesized from the elements. A suspended nanowire is a wire in vacuum chamber held at the extremities. A deposited nanowire is a wire deposited on a surface of different nature: e.g. it could even be a single strip of metallic atoms over a non-conducting surface. A suspended nanowire can be produced by chemical etching of a bigger wire, or bombarding a bigger wire with some highly energetic particles (atoms or molecules).

Another way to produce a suspended nanowire is to indent the tip of an STM in the surface, of a metal near the melting point, and retract it. A common technique for creating a nanowire is the Vapour-Liquid-Solid (VLS) synthesis method. This technique uses as source material either laser ablated particles or a feed gas (such as silane). The source is then exposed to a catalyst. For nanowires, the best catalysts are liquid metal (such as gold) nanoclusters, which can either be purchased in colloidal form and deposited on a substrate or self-assembled from a thin film by dewetting. This process can often produce crystalline nanowires in the case of semiconductor materials.

The source enters these nanoclusters and begins to saturate it. Once supersaturation is reached, the source solidifies and grows outward from the nanocluster. The final product's length can be adjusted by simply turning off the source.

Compound nanowires with super-lattices of alternating materials can be created by switching sources while still in the growth phase. Inorganic nanowires such as $Mo_6S_{9-x}I_x$ (which are alternatively viewed as cluster polymers) are synthesised in a single-step vapour phase reaction at elevated temperature.

Conductivity

The conductivity of a nanowire is expected to be much less than that of the corresponding bulk material. This is due to a variety of reasons. First, there is scattering from the wire boundaries, when the wire width is below the free electron mean free path of the bulk material. In copper, for example, the mean free path is 40 nm. Nanowires less than 40 nm wide will shorten the mean free path to the wire width.

Nanowires also show other peculiar electrical properties due to their size. Unlike carbon nanotubes, whose motion of electrons can fall under the regime of ballistic transport (meaning the electrons can travel freely from one electrode to the other), nanowire conductivity is strongly influenced by edge effects.

The edge effects come from atoms that lay at the nanowire surface and are not fully bonded to neighbouring atoms like the atoms within the bulk of the nanowire.

The unbonded atoms are often a source of defects within the nanowire, and may cause the nanowire to conduct electricity more poorly than the bulk material. As a nanowire shrinks in size, the surface atoms become more numerous compared to the atoms within the nanowire, and edge effects become more important. Furthermore the conductivity can undergo a quantisation in energy: i.e. the energy of the electrons going through a nanowire can assume only discrete values, multiple of the Von Klitzing constant $G = 2e^2/h$ (where e is the charge of the electron and h is Planck's constant).

The conductivity is hence described as the sum of the transport by separate channels of different quantized energy levels. The thinner the wire is, the smaller the number of channels available to the transport of electrons. The conductivity of a nanowire can be studied suspending it between two electrodes. This has been proven by measuring the conductivity of a nanowire while pulling it: while it shrinks, its conductivity decreases in a stepwise fashion and the plateaus correspond to multiples of G.

The quantized conductivity is more pronounced in semiconductors like Si or GaAs than in metals, due to lower electron density and lower effective mass. Quantized conductance can be observed in 25 nm wide silicon fins, resulting in increased threshold voltage.

Use of nanowires

Nanowires still belong to the experimental world of laboratories. However, they may complement or replace carbon nanotubes in some applications. Some early experiments have shown how they can be used to build the next generation of computing devices. To create active electronic elements, the first key step was to chemically dope a semiconductor nanowire. This has already been done to individual nanowires to create p-type and n-type semiconductors.

The next step was to find a way to create a p-n junction, one of the simplest electronic devices. This was achieved in two ways. The first way was to physically cross a p-type wire over an n-type wire. The second method involved chemically doping a single wire with different dopants along the length. This method created a p-n junction with only one wire.

After p-n junctions were built with nanowires, the next logical step was to build logic gates. By connecting several p-n junctions together, researchers have been able to create the basis of all logic circuits: the AND, OR, and NOT gates have all been built from semiconductor nanowire crossings. It's possible that semiconductor nanowire crossings will be important to the future of digital computing. Though there are other uses for nanowires beyond these, the only ones that actually take advantage of physics in the nanometer regime are electronic.

Nanowires are being studied for use as photon ballistic waveguides as interconnects in quantum dot/quantum effect well photon logic arrays. Photons travel inside the tube, electrons travel on the outside shell. When two nanowires acting as photon waveguides cross each other the juncture acts as a quantum dot.

Conducting nanowires offer the possibility of connecting molecular-scale entities in a molecular computer. Dispersions of conducing nanowires in different polymers are being investigated for use as transparent electrodes for flexible flat-screen displays.

Due to their high Young's moduli their use in mechanically enhancing composites is being investigated. Because nanowires appear in bundles, they may be used as tribological additives to improve friction characteristics and reliability of electronic transducers and actuators.

METALLIC NANOANTENNA

A nanoantenna is a device that absorbs small wavelength electromagnetic radiation through resonance. Nanoantennas help to focus light more than is possible with convention materials. Metallic nanostructures may be able to eventually detect single molecules. The bowtie nanoantenna is a structure that is on the order of hundreds of nanometers that can compress light into a region up to 20 nanometers in diameter. The device consists of two triangular pieces of metal, often Gold, with the tips facing each other. By varying the geometeries of the antenna the intensity of various wavelengths of light can be enhanced.

Most materials, including air, water, and plastic, are termed "right-handed" with regard to how light bends when it penetrates the material. Light bends in the opposite direction when it penetrates "left-handed" materials. The nanoantenna is a left-handed type of material that could be used to create a super lens, which helps improve the resolution of medical diagnostic images.

Engineers and scientists demonstrated a plasmonic laser nanoantenna, created by integrating a metallic nanostructure with a commercial semiconductor laser. The optical antenna is able to collect and focus light within a region of tens of nanometers. The optical antenna operates within the visible and infrared portion of the electromagnetic spectrum. It is believed that using this tool with optical storage platforms could result in memory capacities of terabytes.

The technology of solar arrays of nanoantennas is expected to be highly efficient with regard to capturing energy (80% efficiency) and in the process of creating the device (cents per yard). Small square spirals of metal can be printed on plastic sheets. The antennas can capture energy associated with infrared radiation, which reaches the earth from the sun during the day and is emitted by the earth at night.

U.S. DOE's Idaho National Laboratory (INL) has developed a method of producing plastic sheets that contain billions of nanoantennas to convert the sun's infrared rays into electricity. The nanoantennas consist of tiny gold squares or spirals set in a specially treated layer of polyethylene, a common plastic. While the device is not yet practical, it has the potential to run off either sunlight or waste heat.

Engineers at the University of California, San Diego (UCSD) have employed similar "nanowires" to boost the efficiency of organic solar cells, which are made of plastic. Scientists hope to create computers that use photons instead of electrons to transmit signals. Manipulating photons is more difficult than electrons since they do not carry charge. It's theoretically possible to simulate all the properties of electric circuits with plasmonic nanomaterials used to guide light.

Conception of Nanoionics

Nanoionics is the study and application of phenomena, properties, effects and mechanisms of processes connected with fast ion transport (FIT) in all-solid-state nanoscale systems. The topics of interest include fundamental properties of oxide ceramics at nanometer length scales, and fast ion conductor (advanced superionic conductor)/ electronic conductor heterostructures.

Potential applications are in electrochemical devices (electrical double layer devices) for conversion and storage of energy, charge and information. The term and conception of nanoionics (as a new branch of science) were first introduced by A.L.Despotuli and V.I.Nikolaichik (Institute of Microelectronics Technology and High Purity Materials, Russian Academy of Sciences, Chernogolovka) in January 1992.

There are two classes of solid state ionic nanosystems and two fundamentally different nanoionics:

(I) nanosystems based on solids with low ionic conductivity, and

(II) nanosystems based on advanced superionic conductors (alpha-AgI, rubidium silver iodide-family, etc.).

The latter was proposed in. Nanoionics-I and nanoionics-II differ from each other in the design of interfaces. The role of boundaries in nanoionics-I is the creation of conditions for high concentrations of charged defects (vacancies and interstices) in a disordered space-charge layer. But in nanoionics-II, it is necessary to conserve the original highly ionic conductive crystal structures of advanced superionic conductors at ordered (lattice-matched) heteroboundaries.

Nanoionic-I can significantly enhance (up to $\sim 10^8$ times) the 2D-like ion conductivity in nanostructured materials with structural coherence, but it is remaining in $\sim 10^3$ times smaller relatively to 3D ionic conductivity of advanced superionic conductors.

Characteristics of Nanoionics

Being a branch of science and technology, nanoionics is unambiguously defined by its own objects (nanostructures with FIT), subject matter (properties, phenomena, effects, mechanisms of processes, and applications connected with FIT at nano-scale), method (interface design in nanosystems of superionic conductors), and criterion (R/L ~1, where R is the nanosize(s) of device structure, and L is the characteristic length on which the properties, characteristics, and other parameters (connected with FIT) change drastically.

The International Technology Roadmap for Semiconductors, known throughout the world as the ITRS, relates nanoionics-based resistive switching memories to the category of "emerging research devices" ("ionic memory"). The area of close intersection of nanoelectronics and nanoionics can be called nanoelionics. Now, the vision of future nanoelectronics constrained solely by fundamental ultimate limits is being formed in advanced researches.

The ultimate physical limits to computation are very far off from the attained »10^{10} cm^{-2}-10^{10} Hz » region. What kind of logic switches might be used at the near nm- and sub-nm peta-scale integration? The question was the subject matter already in, where the term "nanoelectronics" was not used yet.

Quantum mechanics constrains electronic distinguishable configurations by the tunnelling effect at tera-scale. To overcome 10^{12} cm^{-2} bit density limit, atomic and ion configurations with acharacteristic dimension of L <2 nm should be used in the information domain and materials with an effective mass of information carriers m* considerably larger than electronic ones are required: $m^* = 13\ m_e$ at L =1 nm, $m^* = 53\ m_e$ (L =0,5 nm) and $m^* = 336\ m_e$ (L =0,2 nm). Future short-sized devices may be nanoionic, i.e. based on the fast ion transport at the nanoscale, as it was first stated in.

The examples of nanoionic devices are all-solid-state supercapacitors with fast ion transport at the functional heterojunctions (nanoionic supercapacitors), lithium batteries and fuel cells with nanostructured electrodes, nano-switches with quantized conductivity on the basis of fast ion conductors. These are well compatible with sub-voltage and deep-sub-voltage nanoelectronics and could find wide applications e.g. as in autonomous micro power sources, RFID, MEMS, smartdust, nanomorphic cell, other micro- and nanosystems, or reconfigurable memory cell arrays.

An important case of fast ionic conduction in solid states is that in surface space-charge layer of ionic crystals. Such conduction was first predicted by Kurt Lehovec. A significant role of boundary conditions with respect to ionic conductivity was first experimentally discovered by C.C. Liang who found an anomalously high conduction in the LiI-Al2O3 two-phase system.

Because a space-charge layer with specific properties has nanometer thickness, the effect is directly related to nanoionics (nanoionics-I). The Lehovec effect has become the basis for the creation of a multitude of nanostructured fast ion conductors which are used in modern portable lithium batteries and fuel cells.

Nanosensors

Nanosensors are any biological, chemical, or sugery sensory points used to convey information about nanoparticles to the macroscopic world. Though humans have not yet been able to synthesize nanosensors, predictions for their use mainly include various medicinal purposes and as gateways to building other nanoproducts, such as computer chips that work at the nanoscale and nanorobots. Presently, there are several ways proposed to make nanosensors, including top-down lithography, bottom-up assembly, and molecular self-assembly.

Applications of Nanosensors

Medicinal uses of nanosensors mainly revolve around the potential of nanosensors to accurately identify particular cells or places in the body in need. By measuring changes in volume, concentration, displacement and velocity, gravitational, electrical, and magnetic forces, pressure, or temperature of cells in a body, nanosensors may be able to distinguish between and recognise certain cells, most notably those of cancer, at the molecular level in order to deliver medicine or monitor development to specific places in the body.

In addition, they may be able to detect macroscopic variations from outside the body and communicate these changes to other nanoproducts working within the body.

One example of nanosensors involves using the fluorescence properties of cadmium selenide quantum dots as sensors to uncover tumours within the body. By injecting a body with these quantum dots, a doctor could see where a tumour or cancer cell was by finding the injected quantum dots, an easy process because of their fluorescence.

Developed nanosensor quantum dots would be specifically constructed to find only the particular cell for which the body was at risk. A downside to the cadmium selenide dots, however, is that they are highly toxic to the body.

As a result, researchers are working on developing alternate dots made out of a different, less toxic material while still retaining some of the fluorescence properties. In particular, they have been investigating the particular benefits of zinc sulphide quantum dots which, though they are not quite as fluorescent as cadmium selenide, can be augmented with other metals including manganese and various lanthanide elements. In addition, these newer quantum dots become more fluorescent when they bond to their target cells.

Potential predicted functions may also include sensors used to detect specific DNA in order to recognise explicit genetic defects, especially for individuals at high-risk and implanted sensors that can automatically detect glucose levels for diabetic subjects more simply than current detectors.

DNA can also serve as sacrificial layer for manufacturing CMOS IC, integrating a nanodevice with sensing capabilities. Therefore, using proteomic patterns and new hybrid materials, nanobiosensors can also be used to enable components configured into a hybrid semiconductor substrate as part of the circuit assembly. The development and miniaturisation of nanobiosensors should provide interesting new opportunities.

Other projected products most commonly involve using nanosensors to build smaller integrated circuits, as well as incorporating them into various other commodities made using other forms of nanotechnology for use in a variety of situations including transportation, communication, improvements in structural integrity, and robotics. Nanosensors may also eventually be valuable as more accurate monitors of material states for use in systems where size and weight are constrained, such as in satellites and other aeronautic machines.

Existing Nanosensors

Currently, the most common mass-produced functioning nanosensors exist in the biological world as natural receptors of outside stimulation. For instance, sense of smell, especially in animals in which it is particularly strong, such as dogs, functions using receptors that sense nanosized molecules. Certain plants, too, use nanosensors to detect sunlight; various fish use nanosensors to detect minuscule vibrations in the surrounding water; and many insects detect sex pheromones using nanosensors.

Certain electromagnetic sensors have also been in use in photoelectric systems. These work because the specific sensors called, aptly, photosensors are easily influenced by light of various wavelengths.

The electromagnetic source transfers energy to the photosensors and energizes them into an excited state which causes them to release an electron into a semiconductor. At that point, it is relatively easy to detect the electricity coming from the sensors, and thus easy to know if the sensors are receiving light.

Though more advanced uses of photosensors incorporating other forms of nanotechnology have yet to be implemented into consumer society, most film cameras have used photosensors at the nano size for years.

Traditional film uses a layer of silver ions that become excited by solar energy and clump into groups, as small as four atoms apiece in some cases, that scatter light and appear dark on the frame. Various other types of film can be made using a similar process to detect other specific wavelengths of light, including x-rays, infrared, and ultraviolet.

One of the first working examples of a synthetic nanosensor was built by researchers at the Georgia Institute of Technology in 1999. It involved attaching a single particle onto the end of a carbon nanotube and measuring the vibrational frequency of the nanotube both with and without the particle. The discrepancy between the two frequencies allowed the researchers to measure the mass of the attached particle.

Chemical sensors, too, have been built using nanotubes to detect various properties of gaseous molecules. Carbon nanotubes have been used to sense ionization of gaseous molecules while nanotubes made out of titanium have been employed to detect atmospheric concentrations of hydrogen at the molecular level.

Many of these involve a system by which nanosensors are built to have a specific pocket for another molecule. When that particular molecule, and only that specific molecule, fits into the nanosensor, and light is shone upon the nanosensor, it will reflect different wavelengths of light and, thus, be a different colour.

Production Methods

There are currently several hypothesized ways to produce nanosensors. Top-down lithography is the manner in which most integrated circuits are now made. It involves starting out with a larger block of some material and carving out the desired form. These carved out devices, notably put to use in specific microelectro mechanical systems used as microsensors, generally only reach the micro size, but the most recent of these have begun to incorporate nanosized components.

Another way to produce nanosensors is through the bottom-up method, which involves assembling the sensors out of even more minuscule components, most likely individual atoms or molecules. This would involve moving atoms of a particular substance one by one into particular positions which, though it has been achieved in laboratory tests using tools such as atomic force microscopes, is still a significant difficulty, especially to do en masse, both for logistic reasons as well as economic ones. Most likely, this process would be used mainly for building starter molecules for self-assembling sensors.

The third way, which promises far faster results, involves self-assembly, or "growing" particular nanostructures to be used as sensors. This most often entails one of two types of assembly. The first involves using a piece of some previously created or naturally formed nanostructure and immersing it in free atoms of its own kind.

After a given period, the structure, having an irregular surface that would make it prone to attracting more molecules as a continuation of its current pattern, would capture some of the free atoms and continue to form more of itself to make larger components of nanosensors.

The second type of self-assembly starts with an already complete set of components that would automatically assemble themselves into a finished product. Though this has been so far successful only in assembling computer chips at the micro size, researchers hope to eventually be able to do it at the nanometer size for multiple products, including nanosensors.

Accurately being able to reproduce this effect for a desired sensor in a laboratory would imply that scientists could manufacture nanosensors much more quickly and potentially far more cheaply by letting numerous molecules assemble themselves with little or no outside influence, rather than having to manually assemble each sensor.

Economic Impact

Though nanosensor technology is a relatively new field, global projections for sales of products incorporating nanosensors range from $0.6 billion to $2.7 billion in the next three to four years. They will likely be included in most modern circuitry used in advanced computing systems, since their potential to provide the link between other forms of nanotechnology and the macroscopic world allows developers to fully exploit the potential of nanotechnology to miniaturize computer chips while vastly expanding their storage potential.

First, however, nanosensor developers must overcome the present high costs of production in order to become worthwhile for implementation in consumer products. Additionally, nanosensor reliability is not yet suitable for widespread use, and, because of their scarcity, nanosensors have yet to be marketed and implemented outside of research facilities. Consequently, nanosensors have yet to be made compatible with most consumer technologies for which they have been projected to eventually enhance.

NANOELECTROMECHANICAL SYSTEMS (NEMS)

Nanoelectromechanical systems or NEMS are similar to Microelectro mechanical systems (MEMS) but smaller. They hold promise to improve abilities to measure small displacements and forces at a molecular scale, and are related to nanotechnology and nanomechanics.

There are two approaches most researchers accept as standard paths to NEMS. The top-down approach can be summarised as "a set of tools designed to build a smaller set of tools". For example, a millimeter sized factory that builds micrometer sized factories which in turn can build nanometer sized devices. The other approach is the bottom-up approach, and can be thought of as putting together single atoms or molecules until a desired level of complexity and functionality has been achieved in a device. Such an approach may utilise molecular self-assembly or mimic molecular biology systems.

In 2000, the first Very Large Scale Integration (VLSI) NEMS device was demonstrated by researchers from IBM. Its premise was an array of AFM tips which can heat/sense a deformable substrate in order to function as a memory device.

In 2007, the International Technical Roadmap for Semiconductors (ITRS) contains NEMS Memory as a new entry for the Emerging Research Devices section; an indication that the semiconductor industry is actively considering the technology for implementation in the near (15 years) future. A combination of these approaches may also be used, in which nanoscale molecules are integrated into a top-down framework. One such example is the carbon nanotube nanomotor.

References

Aviram, A., "Molecules for memory, logic, and amplification", *Journal of the American Chemical Society* 110 (17): 5687-5692. doi:10.1021/ja00225a017, 1988.

Cavalcanti, A.; Shirinzadeh, B.; Freitas Jr, Robert A. & Hogg, Tad, "Nanorobot architecture for medical target identification". *Nanotechnology* 19 (1): 015103(15pp), 2008.

Das, S., *et al.*, "Designs for Ultra-Tiny, Special-Purpose Nanoelectronic Circuits", *IEEE Trans. on Circuits and Systems* I 54: 11, 2007.

Frist, W.H., "Health care in the 21st century", *N. Engl. J. Med.* 352 (3): 267-272, 2005.

Goicoechea, J.; Zamarreñoa, C.R.; Matiasa, I.R. & Arregui, F.J., "Minimizing the photobleaching of self-assembled multilayers for sensor applications", *Sensors and Actuators B: Chemical* 126 (1): 41-47, 2007.

Grace, D., "Special Feature: Emerging Technologies", *Medical Product Manufacturing News,* 2008.

LaVan, D.A.; McGuire, Terry & Langer, Robert, "Small-scale systems for in vivo drug delivery", *Nat Biotechnol*, 21 (10): 1184-1191, 2003.

Melosh, N., *et al.*, "Ultrahigh density nanowire lattices and circuits", *Science* 300: 112, 2003.

Patolsky, F.*et al.*, "Detection, stimulation, and inhibition of neuronal signals with high-density nanowire transistor arrays", *Science* 313: 1100-1104, 2006.

Petty, M.C.; Bryce, M.R. & Bloor, D., "An Introduction to Molecular Electronics", *London: Edward Arnold*, 1995.

Postma, Henk W. Ch.; Teepen, Tijs; Yao, Zhen; Grifoni, Milena & Dekker, Cees, "Carbon nanotube single-electron transistors at room temperature", *Science* 293 (5527): 76-79, 2001.

Rutherglen, C. & Burke, P., "Carbon nanotube radio", *Nano Lett.* 7 (11): 3296-3299, 2007.

Bibliography

Akhlesh Lakhtakia (ed), *The Handbook of Nanotechnology*, Nanometer Structures: Theory, Modeling, and Simulation, SPIE Press, Bellingham, WA, USA., 2004.

Buffat, Ph.; Burrel, J.-P., "Size effect on the melting temperature of gold particles", *Physical Review* A 13 (6): 2287-2298, 1976.

Choy JH, Jang ES, Won JH, Chung JH, Jang DJ, and Kim YW., "Hydrothermal route to ZnO nanocoral reefs and nanofibers", *Appl. Phys. Lett.* 84 287, 2004.

Das S, Gates AJ, Abdu HA, Rose GS, Picconatto CA, Ellenbogen JC., "Designs for Ultra-Tiny, Special-Purpose Nanoelectronic Circuits", *IEEE Transactions on Circuits and Systems* I 54 (11): 2528-2540, 2007.

Davis, R. C., *et al.*, "Chemomechanical surface patterning and functionalization of silicon surfaces using an atomic force microscope", *Appl. Phys. Lett.* 82 (5): 808-810, 2003.

Ghalanbor Z, Marashi SA, Ranjbar B., "Nanotechnology helps medicine: nanoscale swimmers and their future applications", *Med Hypotheses* 65 (1): 198-199, 2005.

Hari Singh Nalwa, *Encyclopedia of Nanoscience and Nanotechnology* (10-Volume Set), American Scientific Publishers, 2004.

Hatzor-de Picciotto, A., A. D. Wissner-Gross, G. Lavallee, P. S. Weiss, "Arrays of Cu(2+)-complexed organic clusters grown on gold nano dots". *Journal of Experimental Nanoscience* 2: 3-11, 2007.

Jumana Boussey, Georges Kamarinos, Laurent Montès (editors), Towards Nanotechnology, "Nano et Micro Technologies", *Hermes Sciences Publ.*, Paris, 2003.

Kubik T, Bogunia-Kubik K, Sugisaka M., "Nanotechnology on duty in medical applications", *Curr Pharm Biotechnol*, 6 (1): 17-33, PMID 15727553, 2005.

Levins CG, Schafmeister CE., "The synthesis of curved and linear structures from a minimal set of monomers", *Journal of Organic Chemistry*, 70, p. 9002, 2005.

Lynn E., *Foster: Nanotechnology: Science, Innovation, and Opportunity*, Prentice Hall, December 21 2005.

Michael Rieth and Wolfram Schommers, *Handbook of Theoretical and Computational Nanotechnology* (10-Volume Set), American Scientific Publishers, 2006.

Narayan RJ, Kumta PN, Sfeir C, Lee D-H, Olton D, Choi D., "Nanostructured Ceramics in Medical Devices: Applications and Prospects", JOM 56 (10): 38-43, 2004.

Shetty RC., "Potential pitfalls of nanotechnology in its applications to medicine: immune incompatibility of nanodevices", *Med Hypotheses* 65 (5): 998-9, 2005.

Sundrani D, Darling SB, Sibener SJ, "Hierarchical assembly and compliance of aligned nanoscale polymer cylinders in confinement", *Langmuir 20* (12): 5091-9, June 2004.

Taniguchi, N., "On the Basic Concept of 'Nano-Technology'," Proc. Intl. Conf. Prod. London, Part II, *British Society of Precision Engineering*, 1974.

William Sims Bainbridge, *Nanoconvergence: The Unity of Nanoscience, Biotechnology, Information Technology and Cognitive Science*, Prentice Hall, June 27 2007.

Zsigmondy, R., "Colloids and the Ultramicroscope", *J. Wiley and Sons*, NY, 1914.

Index